The World Zinc Industry

The World Zinc Industry

Satyadev Gupta
The University of the West Indies

LexingtonBooks
D.C. Heath and Company
Lexington, Massachusetts
Toronto

Library of Congress Cataloging in Publication Data

Gupta, Satyadev.
The world zinc industry.

Bibliography: p.
Includes index.
1. Zinc industry and trade. 2. Title.
HD9539.Z5G86 338.2'7452 81-47075
ISBN 0-669-04587-x AACR2

Published simultaneously in Canada

Printed in the United States of America

International Standard Book Number: 0-669-04587-x

Library of Congress Catalog Card Number: 81-47075

To
My Father

Contents

List of Figures

List of Tables

Acknowledgments

I wish to express my deepest gratitude to Professors J.R. Williams, F.T. Denton, and R.A. Muller for their continuous help and inspiration at all stages of this book's preparation. I thank Professors F.G. Adams and J.R. Behrman, in the same spirit, for their valuable help. And I thank Dr. A.V. Cammarotta, Jr., and Dr. W.Y. Mo of the U.S. Bureau of Mines for helpful discussions on the subject.

Many institutions provided financial and other support for this study. I am especially grateful to McMaster University for a generous fellowship, and the Universities of Delhi and the West Indies for study leave and grants. I also wish to acknowledge my gratitude to the University of Pennsylvania for providing various facilities during my stay as a visiting research scholar.

I would like to take this opportunity to express my gratitude to my teachers, Professors A. Bhaduri, J. Bhagwati, S. Chakravarti, M. Duttachaudhari, K.L. Krishna, A. Kubursi, A.L. Nagar, P.K. Pattanaik, A.K. Sen, and D.M. Winch for their contributions and guidance in my education. I am also thankful to Mr. W. Dookeran, Dr. E.B.A. St. Cyr, and other colleagues at the University of the West Indies for their encouragement during the preparation of this book.

I offer my deep appreciation to my wife, Sarita, and daughter, Hansa, and to my parents and other members of my family for their patience, support, and encouragement. Finally, I am grateful to Mrs. V. Youssef for suggesting many linguistic improvements, and to Mrs. J. Prempeh, Miss A. Rampersad, and Miss Ava Seaton for admirable typing assistance. Nonetheless, I alone bear the responsibility for any shortcomings that may remain in the study.

The World Zinc Industry

1 Introduction and Outline

Introductory Remarks

The recent worldwide inflation and the severe recession of the postwar period may be attributed, in part, to the abrupt rise in prices of primary commodities. Minerals, metals, and other industrial raw materials seem to have played a leading role. For instance, during the three-year period from 1972 to 1974, prices of some metals (aluminum, copper, tin, and lead, for example) rose by more than 200 percent, not to mention the more familiar rise in the price of oil and some other minerals. The price of zinc, one of the widely used industrial raw materials, increased by more than 300 percent.[1] A similar surge of prices has been registered for many other primary commodities, including agricultural goods.[2] This rise in the prices of primary commodities, and particularly of industrial raw materials, may have contributed significantly to the so-called cost-push inflationary pressures in the industrial sectors. These inflationary pressures together with some concurrent recessionary forces may have resulted in the high levels of unemployment and prices in the industrialized world. This rise in prices in the industrialized countries is transmitted back to the less-developed countries where, in turn, it is fed back in the inflationary process through the same primary commodities, resulting in the subsequent increases in prices.[3]

This worldwide instability in the growth of free-market economies has attracted the attention of many economists and has encouraged them to have a closer look into the structure, behavior, and performance of the primary commodity markets.[4] Zinc somehow seems to have escaped the attention of the economists altogether. The reasons for neglect of the zinc industry are not hard to find, however. *First*, the apparent structure of the zinc industry does not provide an obvious indication of organizational behavior for its modeling. *Second*, although zinc is a very widely used raw material in many industries such as construction (galvanizing steel, alloyed with copper to produce brass, in the form of rolled zinc sheets for roofs, and so forth), automobiles (used as die casts), rubber, chemicals, printing, armaments (in the form of brass), agriculture, and nutrition, the use of zinc is very thinly distributed, so that it accounts for only a very small proportion of the total cost in any given final product. This probably makes zinc less attractive to the economists as compared to the other commodities which account for larger shares of production costs. *Third*, as evidenced by the experience of the author, the structural parameters of the zinc market are very sensitive to specification errors in an econometric study. *Fourth*, except for Mexico, Peru, and some smaller producers, zinc is produced in relatively rich countries where foreign exchange earnings from zinc do not form a substantial proportion of the balance of payments. Thus, a source of motivation that played an important role in the study of many primary commodities in the late 1960s and the early 1970s is absent in the case of zinc.

Nevertheless, the recent rise in the price of zinc relative to other primary commodities has been so alarming that it warrants a thorough study in terms of market structure, behavior, and performance.[5] Furthermore, such a study is important for the policymakers and planners of the zinc industry.

Basic Aims and Overview of the Study

The basic aims of the study are (1) to investigate the market structure of the zinc industry, (2) to build an econometric model for the industry that may serve as an instrument for policy formulation, and (3) to study the behavior and performance of the industry by the use of this model.

The exact nature of the organizational structure of the world zinc industry is not apparent from the general literature available. However, for realistic model-building, an understanding of its nature is indispensable.[6] From the data available it is possible to trace the organizational structure of the industry in terms of both (1) countries as the units of control, and (2) corporate groups as the units of control in the market for zinc. Since, in free market economies (FME), it is the units of financial control—the corporate groups—that are more important, more emphasis is given to this aspect. Given the number and size of the corporate groups operating in the world zinc industry, their multinational operations are also investigated to provide some additional information with regard to their financial control. As is typical of most mineral industries, zinc ore produced by the primary producers must undergo further processing (smelting) before it can be used by industrial consumers. Integration of the production processes (vertical integration) yields further market power if the ore producers and smelters are integrated. This aspect has been investigated for present and future possibilities of vertical integration. Coordination of the market by international agencies and intervention in the market through national policies have also been discussed.

A thorough study of the zinc market reveals (1) that producers in the United States exercise monopoly power in the market, and (2) that the industry outside the United States is more likely to follow the rules of competitive behavior. These elements have been incorporated into the specification of the econometric model of the industry. Various other institutional and technological considerations are also taken into account in the detailed design of the model.

On the demand side, the model is disaggregated into seven major zinc-consuming regions. The classification of areas depends on their shares in the market, their stages of economic development, and their traditional preferences for zinc. In the second version of the model, the demand side is further disaggregated according to six final sectors of demand, to account for differences in technology and response to prices in the different final demand categories. Both secondary supply and primary supply have been considered explicitly in the system. Differences in the operational costs and the age of mines in different countries are incorporated through the disaggregation of the world supply into major producing regions. Both the stock and the flow-adjustment mechanisms together with appropriate lagged responses in the variables determine price behavior in the world zinc industry. Both the competitive market and the U.S. market are linked through prices, interregional trade, and exchange rates. The model could, therefore, be linked to larger country models to evaluate the influence of certain policies. Both versions of the econometric model of the zinc industry have been estimated by appropriate techniques, as discussed later.

The resultant estimated versions of the model are subjected to testing by dynamic simulation to determine their predictive ability. Both versions perform well, which gives us some confidence in their use for policy evaluations.[7]

As an experiment, a set of six policies or market scenarios, is considered and studied by means of multiplier simulation techniques appropriate for nonlinear, dynamic, simultaneous systems. More specifically, the probable performance of the industry is investigated for situations involving exogenous changes in economic activity, in technology in the consumer industries, in the prices of the substitute materials, and in U.S. government policies for the protection of the domestic industry and the stabilization of the prices in the world zinc market.

Organization of the Study

The study has been organized along the lines discussed above. In chapter 2 the salient features of the international market for zinc have been studied in detail. In the first half of the chapter the supply aspects including reserves, resources, secondary zinc (zinc recovered from scrap), and concentration in zinc market by producer countries, as well as demand aspects relating to consumption structure of zinc, substitution possibilities, concentration on the demand side by countries, and likely future developments of the above elements of the market have been investigated. In the latter half of the chapter aspects of international trade in zinc, the price system including both the structure and behavior of prices in the zinc market, and national policies influencing the international market have been studied. An appendix to this chapter deals with technological aspects of consumption and production of zinc.

Chapter 3 investigates the organizational structure of the world zinc market in terms of the units of financial control. It also explores the possibility of links between the various corporate groups, their multinational operations, and the question of vertical integration. In the last part of the chapter the role of the various international organizations in coordinating the world zinc industry is studied. In the appendix to this chapter guidance is sought in understanding both the recent past and prospects for the future from the history of cartelization during the inter-world-war period.

Chapter 4 surveys the attempts of some economists at model-building for some mineral commodities, and looks into aspects of mineral economics and its likely influence on model-building. At the end of the chapter, we provide a general sketch of our own model of the world zinc industry.

Chapter 5 deals with the specification of the two versions of the model in detail, the methodology of estimation, and the analysis of the structural parameters estimated.

In chapter 6, the validity of our econometric models is tested through the techniques of dynamic simulation. The models then are used, in the next chapter, for a set of six policy simulations. In particular, behavior patterns of the industry in response to the fluctuations in business activity, changes in technology, and changes in the prices of substitutes are investigated. So, too, is the influence of the U.S. strategic stockpile policy. Chapter 8 concludes the study with some suggestions for further research in this area.

Notes

1. In order of the quantities consumed and produced of the various nonferrous metals, zinc occupies the third place, aluminum and copper occupying the first and the second place, respectively.

2. See Adams and Behrman (1977).

3. The theoretical models that would incorporate the micro-behavior of primary commodities explicitly into the macro-models are too scarce to find. However, some attempts have begun to be made in terms of applied economic analysis. For example, see Adams (1973a).

4. Labys, ed. (1975), in his extensive bibliography, lists about 241 studies of commodity markets, ranging from aluminum to wool, a majority of which were undertaken during the last ten years.

5. Zinc falls into the category of strategic materials of the U.S. government, and in the list of important commodities of the United Nations. This is also reflected in the numerous attempts of the U.S. government to intervene in the working of the world zinc market, and that of the UN to coordinate policymaking in the industry. For details, see chapters 2 and 3.

6. The proposition is debatable. For example, Friedman and many other economists would insist only on the predictive power of a model, whereas a large number of economists such as Koopmans would insist on both the realism of the postulates as well as the predictive power of the theory derived from the postulates. We take the latter approach in this study. For an extended discussion of these aspects, see Friedman (1953, p. 41), and Koopmans (1957, p. 138).

7. See plots of dynamic simulations in the appendix.

2 The International Market for Zinc

Zinc is an internationally traded intermediate input widely used in many industries. Industrialized countries are, therefore, major consumers of zinc. Mine production of zinc is scattered generally throughout the world, though more concentrated in Australia, Canada, Mexico, Peru, the United States, and the USSR. The first four of these countries share more than 75 percent of the world exports of raw zinc. The major importers are the United States, Japan, and the European countries. These same countries are also the major producers and consumers of zinc metal. Over the last few years, however, vertical integration in the raw zinc producing countries is increasing, thus probably changing the balance of market power in their favor.

The U.S. policies, such as quotas, tariffs, subsidies, and the strategic stockpile program, in the past, had the effect of dichotomizing the world zinc market. Nevertheless, both the markets were linked through international trade—although to a limited extent—in both raw zinc and zinc metal. The world zinc market outside the United States, with the London Metal Exchange (LME) as its apex, although competitive, has been relatively unstable. Extreme instability of zinc prices at the LME during the years 1963–1964 induced the major producers outside the United States to introduce a fixed price of zinc, called the European Producers Price (EPP) for transactions outside the United States. As a result, the turnover of zinc at the LME had decreased to about 10 percent of the total world trade in zinc by 1975. Since 1971, transactions with the United States have also been included under the EPP system.

For an adequate quantitative analysis, however, a detailed study of the various institutional, technological, and behavioral aspects of an industry is deemed necessary. In view of an absence of a systematic study of these aspects of the world zinc industry, chapters 2 and 3 are devoted to this objective. In the following sections of this chapter, supply-demand aspects, the price mechanism, and the national and international policies affecting the international market for zinc are investigated. Technological aspects are discussed in the appendix to this chapter.

World Supply of Zinc

The supply of zinc, like any other metal, depends on primary and secondary resources, technological developments in exploration, mining and smelting operations, and economic environments.[1] Primary resources are distinguished from secondary resources, as the former implies availability of the mineral in nature, whereas the latter refers to the residue from completion of the processes of fabrication and/or consumption. Other things being equal, a technological development in exploration can increase the resources available to the world. An improvement in mining, milling, and smelting technologies may reduce the cost and/or improve the recovery of the metal, thus increasing supply at the given prices.

Reserves and Resources

Geographical distribution of mineral resources containing the metal is very important for the long-run analysis of both the organizational structure and behavior of the world market. Relatively high concentration of a mineral in a small geographical region may yield a monopolistic power to the producers of the mineral, as recently observed in the cases of oil, bauxite, tin and copper.[2] If the mineral in question is thinly distributed, a competitive market structure and the corresponding market behavior is more probable. However, even if the mineral is thinly distributed, the possibility of competitive behavior may be limited through concentrated control, either due to political interference and/or due to a concentrated ownership structure, such as the existence of a very few multinational corporations controlling the resources. Here, the discussion is limited to the geographical distribution only, leaving a detailed examination of the other institutional aspects for the next chapter.

In 1975, the total measured and indicated world zinc reserves at the current prices were estimated at 149 million (mn.) short tons (s.t.). If the other "inferred" and "hypothetical" (but not "demonstrated"), economic resources in the known areas are added, the figure is increased to 270 mn. s.t. In addition, educated guesses have suggested that more than 5,000 mn. s.t. of zinc exist in undiscovered and subeconomic (at the current prices) resources. It is believed that about 34 percent and 25 percent of these resources probably lie in Europe and America, respectively. About 14 percent may be obtained from the seabed, and the rest are distributed in Asia, Africa, and Oceania (see table 2-1). At this stage, it is not possible to obtain any reliable estimates of breakdown in terms of countries. We do have, however, some reliable estimates of a breakdown by country of the resources that are considered economically viable at current economic conditions. America alone shares about 50 percent of the total world resources, followed by Europe (24 percent) and Australia (13 percent). By country, Canada (55 mn. s.t.) leads the list followed by the United States (50 mn. s.t.), Australia (36 mn. s.t.), and the USSR (24 mn. s.t.). Excluding the United States, the USSR, Japan, and some other European countries, whose requirements of

Table 2–1
World Zinc Resources, 1975
(million short tons zinc content)

Area/ Country	*Reserves*[a]	*Other*[b]	*Total*	*Total including Undiscovered and Subeconomic Resources*
North America	69.0	44.0	113.0	1,100
United States	30.0	20.0	50.0	
Canada	34.0	21.0	55.0	
Mexico	4.3	1.7	6.0	
Central America	0.7	1.3	2.0	
South America	9.0	9.0	18.0	300
Peru	3.4	2.6	6.0	
Brazil	3.3	3.7	7.0	
Other	2.3	2.7	5.0	
Europe	34.0	30.0	64.0	1,900
Ireland	8.0	4.0	12.0	
Poland	2.5	4.5	7.0	
USSR	12.0	12.0	24.0	
Yugoslavia	3.6	2.4	6.0	
Other	7.9	7.1	15.0	
Africa	7.0	8.0	15.0	300
Zaire	2.0	2.0	4.0	
Zambia	1.0	1.0	2.0	
Other	4.0	5.0	9.0	
Asia	12.0	12.0	24.0	750
People's Republic of China	1.2	3.8	5.0	
India	2.0	1.0	3.0	
Japan	5.0	3.0	8.0	
North Korea	1.6	1.4	3.0	
Other	2.2	2.8	5.0	
Oceania	18.0	18.0	36.0	450
Australia	18.0	18.0	36.0	
				Seabed: 800
World total	149.0	121.0	270.0	5,600

Source: V. A. Cammarota, H. R. Babitzke, and J. M. Hague, "Zinc, Mineral Facts and Problems," Bulletin 667 (Washington, D. C.: U.S. Bureau of Mines, 1975). Reprinted with permission.

[a]"Measured + Indicated," or "Demonstrated."

[b]Inferred reserves and hypothetical economic resources in known areas.

the mineral are likely to be in excess of their resources, Canada, Australia, and Ireland together possess more than two-thirds of the resources. Thus, in the future, given the present estimates of the resources and the economic conditions, these three countries together will dominate the world supply of raw zinc.

It should be emphasized that the above estimates of resources are conditional on the current state of economic, scientific, and technological development. A rise in the relative price of zinc may make hitherto subeconomic resources economically viable. An advancement in the science of exploration and in techniques of mining, milling, and smelting may increase the available resources and the metal recovery from the available resources. In fact, resources of any mineral may best be considered as a working inventory at a particular level of economic and technological development. It may not, therefore, be surprising that the estimates of resources over time have changed frequently. For example, in the late 1940s, world zinc reserves were estimated at 77 mn. s.t.; whereas, the cumulative mine production of zinc during 1951–1974 passed 106 mn. s.t. and there exists a very large outstanding stock of reserves, as noted above.

Mine Production of Zinc

Usually, explorations for a mineral are undertaken after a long-term tendency of increased requirements has been observed. Exploration and subsequent development of a mine to production stage takes about eight to ten years. In the short term, however, mine production, given the stock of resources, depends on technological capacity of the mines and some other economic factors such as expected prices, wages, prices of coproducts, and metal stocks. An increase in metal prices may bring marginal mines into production, utilize the excess mine capacity, or increase the mine capacity with a lag of one or two years. However, adverse economic factors do not imply an immediate closure of a mine. Closing a mine due to a recession and opening up again later is a very costly business. Usually, in the short term, mines continue to operate, even at much less than capacity level and despite average variable costs which may be higher than the average revenue. Mine closures usually take place only if the sum of average revenue and average cost of reopening the mine is lower than the average variable costs.[3]

World zinc mine production during the 1955–1974 period has more than doubled and amounted to close to 5.8 mn. metric tons (m.t.) in 1974. This is comparable with an increase of 150 percent in lead, about 220 percent in copper, and about 750 percent in bauxite during the same period. The annual compound growth rate of the free market economy (FME) world zinc mine production for 1960–1974 was over 4 percent. However, the growth rates in different countries have varied substantially (see table 2-2). For instance, closure of many old mines in the United States, the leading producer of zinc ore in the world until the late fifties, has resulted in only a 1 percent growth rate during the 1960–1974 period. Canada, on the other hand, the leading producer of zinc ore since the early sixties, has observed a growth rate close to 9 percent. This high growth rate in Canada is largely attributable to the opening of many new mines in the province of Ontario. A similar difference in growth rates was observed in Mexico and Peru, the two other large mine producers of zinc, their growth rates being 0.5 and 6 percent, respectively. Mexico, which produced about 70,000 metric tons more than did Peru in 1960, was surpassed by Peru by about 140,000 tons in 1974. Australia, the third largest producer of zinc ore in the free market world, and all the European countries together experienced a growth rate of about 3 percent.[4] Japan, Zaire, and Zambia have also been important producers. However, the shares of these countries have been declining, while some countries such as Argentina, Bolivia, Iran, South Korea, and Southwest Africa, although insignificant in terms of their shares in comparison to the world zinc market at present, have shown important prospects for increasing mine production of zinc in the FME world.

The major producers of zinc ore in the centrally planned economies (CPE) are the USSR (the second largest producer in the world), Poland, Bulgaria, the People's Republic of China, and North Korea. Although production of zinc ore in the USSR, China, and North Korea in the past has increased substantially, the production of zinc ore in Poland and Bulgaria has

Table 2–2
FME World Zinc Ore Production
(thousand metric tons, zinc in concentrates)

Area or Country	*1960*	*Percentage of Total 1960*	*1974*	*Percentage of Total 1974*	*Annual Compound Growth Rate 1960–1974*
Australia	294.8	11.5	429.0	9.7	2.7
Canada	390.0	15.3	1,237.3	27.9	8.6
Mexico	243.6	9.5	262.0	5.9	0.5
Peru	178.0	7.0	397.2	8.9	5.9
United States	431.1	16.8	498.3	11.2	1.0
Europe	534.8	20.9	818.4	18.5	
Rest of world	485.8	19.0	795.4	17.9	
Total	2,558.2	100.0	4,437.6	100.0	4.0

Source: "Zinc," *Mineral Bulletin*, MR 159 (Ottawa: Department of Energy, Mines and Resources, 1976), p. 32. Reprinted with permission.

remained more or less constant. The share of the CPE world in the total world mine production of zinc during 1956–1974 has increased from 17 percent to 27 percent. However, the consumption of zinc in the CPE world has also increased from 17 percent to 26 percent during the same period, resulting in a very limited flow of zinc between the FME and the CPE countries of the world.

Secondary Supply

Besides the primary resources of supply, secondary sources, the so-called surface mines, and new scrap have been increasingly contributing to the total supply of metals. "Surface mines" refers to the accumulated pile of those fabricated materials containing metals which are worn out or discarded. In the case of zinc, recovery of the metal is possible only in a few areas of its consumption. Particularly, zinc used in galvanizing and chemicals is not recoverable. These two uses of zinc alone constitute about 40 percent of the total zinc consumed. Zinc is recoverable mainly from the zinc-base alloys such as brass and die-cast alloys. The principal zinc-bearing scrap materials include automobile parts, home appliances, roofing sheets, and so forth. On the average, zinc recovered from this old scrap has accounted for about 5 percent of the total zinc supply.[5]

New scrap consists of wastes and surplus material left over as residual in the process of making semi-manufactured or fully-manufactured goods. It is collected at the galvanizers, die-casters' plants, and at the plants of some manufactured goods containing zinc in substantial quantities. The collected scrap is returned to the refineries for remelting and is used subsequently in further manufacturing. The proportion of zinc recovered from new scrap at present amounts to about 5 percent of the total zinc consumed in a given year.[6]

Smelter Production

Historically, geographical distribution of zinc metal production has been very different from that of the mine production of zinc. During the early decades of this century, more than 90 percent of the zinc metal was produced in some of the European countries and the United States. While the U.S. market was protected by high tariff walls, European producers attempted to dominate the rest of the world zinc market by cartel-like actions. To this end, a formal European zinc cartel was formed and broken several times from 1885 to 1935.[7] This concentration of metal production in these regions has continued until recently. In 1960, the share of some of the major European countries and the United States in the FME world zinc metal production was 70 percent, whereas these countries produced only 36 percent of the FME world zinc mine production.[8]

The main reasons for this historical concentration of metal production in the few countries mentioned lay in cheaper sources of energy, technological supremacy in the science of metallurgy, and easy accessibility of the lumpy investments required for the smelting plants. The countries with a larger production of zinc ore lacked both coal and hydroelectric processes. Smelting technology often requires a large capital investment to start production at an efficient level, which investment was not easily available in these countries. Further, developments in technology were often kept secret so as to avoid the possibility of smelting in the countries that were rich in minerals containing zinc.[9]

During the fifteen years between 1960 and 1974, the production of zinc metal in the FME world has increased at the rate of 4.2 percent per year (see table 2–3 below). However, due to the closure of many smelting plants, the United States has lost its leading position to Japan, which has emerged as the largest producer of zinc metal. The closure of many plants in the United States during the 1969–1973 period (see table 2-4) occurred because of their inability to meet the requirements of the recent environmental legislation instituted in that country.[10] Similar legislation has also been passed in some other countries, notably Japan and some European countries. But in these areas, the closed plants have been replaced by the new ones which meet the present legislative requirements.

Table 2–3
FME World Zinc Metal Production
(thousand metric tons)

Area or Country	*1960*	*Percentage of Total 1960*	*1974*	*Percentage of Total 1974*	*Annual Compound Growth Rate 1960–1974*
Australia	122.2	5.0	283.8	6.5	6.2
Canada	236.7	9.7	437.7	10.0	4.5
Mexico	52.9	2.2	137.0	3.2	7.0
Peru	32.5	1.3	70.7	1.6	5.7
United States	791.5	32.5	574.9	13.2	(−2.3)
Europe	922.5	37.8	1,708.7	39.2	4.5
Japan	180.5	7.4	850.0	19.5	11.06
Rest of world	100.4	4.0	299.3	7.0	7.8
Total	2,439.2	100.0	4,362.1	100.0	4.2

Source: "Zinc," *Mineral Bulletin*, MR 159 (Ottawa: Department of Energy, Mines and Resources, 1976), p. 33. Reprinted with permission.

Table 2–4
Closures of U.S. Zinc Smelters, 1969–1973

Company	*Location*	*Type of Smelter*[a]	*Annual Capacity (Short tons)*	*Year Closed*
Anaconda	Blackwell, Oklahoma	E.	90,000	1969
Eagle-Pitcher	Henrietta, Oklahoma	H.R.	55,000	1969
American Zinc Co.	Dumas, Texas	H.R.	58,000	1971
American Zinc Co.[b]	Saugat, Illinois	E.	84,000	1971
Mathiesen & Hegeler	Meadowbrook, W. Va.	V.R.	45,000	1971
New Jersey Zinc Co.	Depue, Illinois	V.R.	65,000	1971
Anaconda	Great Falls, Montana	E.	162,000	1972
Amax	Blackwell, Oklahoma	H.R.	88,000	1973
Asarco[c]	Amarillo, Texas	H.R.	53,000	1973
			616,000[d]	

Source: *Engineering & Mining Journal*, various issues.

[a]H.R.: Horizontal Retort; V.R.: Vertical Retort; E.: Electrolytic

[b]Amax has purchased the Saugat plant, modernized and reopened it in 1975.

[c]Asarco's Amarillo, Texas Plant was allowed to operate until December 1973 with a condition to comply with the Texas Air Control Board's standards. Asarco has appealed to extend operations through 1975.

[d]Excluding reopened Saugat plant.

Recently, with the removal of initial barriers to setting up smelting plants and the recognition of the benefits of employment and value added through integrating zinc ore and metal production within the same country, zinc-ore-producing countries have been increasingly trying to smelt the ores within their own territory. The major contribution to the high growth rates in zinc metal production have come from the zinc-ore-producing countries (except for Japan). Australia, Canada, Mexico, and Peru together have thus increased their share from 18 to 21 percent.

The recent call for the New International Economic Order at the United Nations also proposes that all the raw materials including minerals be processed within the raw material producing country itself.[11] This is likely to have some impact on the tendency in this direction already observed in the past. Although the data on future developments in this direction are not sufficiently available, we do have some information based on the declared plans for building up new plants or expanding the existing ones in some countries during the period 1975–1980. The smelter capacity in the FME world is expected to increase from about 5 million metric tons (m.t.) in 1974 to about 6.4 million m.t. in 1980, an increase of about 30 per cent. The share of the EEC countries and Japan, who have produced only 14 percent of the zinc ore in the FME world, will decline from 50 percent in 1974 to 40 percent in 1980.[12] In fact, no new plant construction is expected in any of these countries during this period. The United States is the only major country among those dependent on foreign ores where smelting capacity is likely to increase (from 635,000 m.t. to 859,000 m.t.). This, however, reflects only a replacement of some of the smelter closures during the 1969–1973 period. Among the European countries, major capacity expansion is expected to be undertaken in Finland, Yugoslavia, and Spain. Whereas Finland and Yugoslavia will just manage to break even with their mine and smelter production, Spain may have to depend largely on imported zinc concentrates for its expanded capacity.

Of the major zinc producers of the FME world, only Mexico and Peru plan to increase their smelter capacity to a significant degree. With the planned expansions of smelter capacity, almost three-fourths of the zinc mine industry in these two countries will be vertically integrated. In Australia, where 70 percent of the zinc mine production is already vertically integrated, there do not seem to be any plans for further increase in smelting capacity at present. Canada plans to increase her smelting capacity from 557,000 m.t. in 1974 to 786,000 m.t. in 1980. Even by 1980, however, about 50 percent of the Canadian mine production of zinc will have to be exported to zinc reduction plants outside the country. A further increase in the smelting capacity in Australia and Canada, the largest producers of zinc ore, may in the future change the structure of the world zinc industry to a great extent. A large part of the world market in zinc concentrates, as a consequence, may be eliminated; a number of the smelters in the EEC, Japan, and the United States may have to be closed down; and, more importantly, the balance of power in the world zinc market may turn in favor of the major zinc ore producing countries.

World Demand for Zinc

In terms of consumption, zinc ranks third (behind copper and aluminum only) among all the nonferrous metals in the world.[13] The construction and manufacturing (particularly automobiles) industries are the major users of zinc. In the United States in 1974, where the data by the sector of final demand are available, the construction industry used about 38 percent of the zinc metal, followed by the transportation industry (27 percent), the electrical industry (13 percent), and machinery (13 percent). The rest was used by the various other industries such as household appliances (non-electrical), lithography, batteries, artistic goods, and so forth.[14] In addition to these direct uses, zinc compounds made either from zinc metal or zinc ore have found a wide variety of applications in industries such as rubber (mainly automobiles), paint, photocopy, pharmaceuticals, cosmetics, and nutrition. In most instances, however, unlike copper and aluminum, both the quantity and the cost factor of zinc in the end product is too small for an immediate recognition of its importance.

Continuous research and development for extending the uses of zinc and the corresponding industrial growth have resulted in a significant increase in the consumption of zinc during the last century. Since 1900 consumption of zinc has increased from 600,000 to 6 million tons. During the period 1960–1974, consumption of zinc grew at an annual rate of 4.6 percent. In per-capita terms, this amounts to an increase from 1.3 kg in 1960 (as compared to 5 kg for all the nonferrous metals) to 1.8 kg in 1974.

Growth in consumption has been, however, very different between countries, due mainly to their differing stages of industrial development. For instance, from 1960–1974 growth rate in the United States and the EEC was recorded at only about 3 percent, as compared to some other countries where growth rates ranged between 5 and 15 percent. Canada and the European countries (other than the EEC) observed growth rates of about 7 percent. Japan has made the largest contribution (in total) to the consumption of zinc, at a growth rate of 9 percent. Growth rate in consumption of zinc in Australia, however, was exceptionally low, at 1.9 percent, and in the developing countries has varied between 7 to 17 percent. Thus, in the future, the consumption of zinc may be expected to grow with the increasing industrialization of the less-developed parts of the world.

Concentration in the Consumption of Zinc

The consumption of zinc, as observed above, is concentrated in the relatively more industrialized countries of the world. The United States, the EEC countries, and Japan together accounted for about 80 percent of the FME world zinc consumption in 1960. Their share, however, decreased to 72 percent in 1974. In this group of countries, whereas Japan during this period doubled its share (from 7.7 to 14.8 percent), the share of the United States and the EEC countries declined from 32 and 40 percent to 27 and 31 percent respectively. The United States even at present, however, occupies the leading position, followed by Japan (14.5 percent), West Germany (8.5 percent), France (6.7 percent), and the United Kingdom (5.9 percent).[15] The shares of the other developed countries (excluding Australia, whose share declined from 3.2 to 2.6 percent) and all the developing countries in the FME world increased from 10 and 7 percent to 13 and 12.5 percent, respectively.

The differentials in the increments for the consumption of zinc in the various countries very much agree with the variations in the increments for industrial production in these countries. For instance, the increases in the industrial production between 1963 and 1974 (1963 = 100) were recorded at 164 for the United States, 320 for Japan, 170 for West Germany, 185 for France, 135 for the United Kingdom, 164 for the rest of the developed world (FME world only), and 220 for all the developing countries together in the FME world. Further, in the short run as well the consumption of zinc may be seen to vary according to the cyclical fluctuations in industrial production in the various countries.

Consumption Structure of Zinc

Although the total consumption of zinc varies according to industrial production in general, variations in some of the end-use categories of zinc depend more appropriately on some particular industries alone. Further, the consumption structure of zinc differs substantially between countries. Thus, an appropriate identification of demand structure requires a closer look at the structure of consumption of zinc in different countries, both for analysis and for estimation of demand functions.

A more appropriate classification of the total consumption of zinc for economic analysis would require the division of zinc consumption into categories according to the sectors of final demand. Data for such a division are not available, either in the aggregate for all the countries, or for different countries separately, except for the United States. The classification conventionally followed in the publications on the consumption of zinc is, rather, in terms of the "intermediate use"—conventionally called the "end-use" categories. The consumption of zinc is usually divided in terms of six end-use categories: galvanizing, die casting, brass, rolled zinc, zinc oxide, and miscellaneous. However, fortunately, one may easily identify some of the major categories with some broad sectors of final demand. This will allow us to use the end-use classification, as available, in quite a meaningful way for the study of policy problems and business fluctuations arising from the movements in some of the major final demand variables. This link will be mentioned briefly here, leaving a detailed discussion of a technical nature to the appendix of this chapter.

Zinc's anticorrosion property, together with its lower melting point, its solubility in copper and some other metals, its inherent ductility, and its maleability are the major characteristics responsible for its use in galvanizing steel, in alloys for die casts, in brass, in the form of rolled zinc, and zinc oxide. Galvanized steel products are used primarily in construction, although their use is also increasing in some home appliances, office equipment, and automobile underbody parts. The automobile industry consumes more than half of the die-casts production, the rest being used in many household appliances, office equipment, and so forth. Parts made of brass (zinc-copper alloy), rolled zinc, and zinc oxide are also used in construction, such as roofing materials, gutters, hardware, and paints. The chemical and rubber industries are the major users of zinc oxide.

Galvanizing and die casting are the major uses of zinc and together account for more than 60 percent of the total consumption of zinc. Brass is the next major use, accounting for about 17 percent, the rest being distributed for use in the form of rolled zinc, zinc oxide and many other miscellaneous uses.[16]

This pattern of zinc consumption, however, is not uniform in the various countries. In general, differences in the consumption pattern arise because of differences in the stages of development, industrial structure, and consumer preferences. For instance, the use of galvanized products varies from about 55 percent in Japan to only 26 percent in the United Kingdom. The European countries have some traditional preference for using brass rather than galvanized steel in some constructional applications, as compared to other countries. Share of zinc consumption for die casts varies among the countries because of differences in the proportion of automobile production in total industrial production, as well as the preferences for using zinc die casts in automobiles. Thus, the share of zinc die casts in the total consumption of zinc ranges from 20 percent in the United Kingdom to 35 percent in the United States. For similar reasons, the proportion of rolled zinc used in construction in some of the European countries differs substantially.[17]

Substitution Possibilities

Zinc is often used both as a complementary and a substitute material with other ferrous and nonferrous metals. Steel, aluminum, and copper are the main complementary metals for the use of zinc in galvanizing, die casting, and brass, respectively. For every 1,000 metric tons of steel produced, about 4 metric tons of zinc are used for galvanizing. About 4 percent of aluminum is used for zinc die-cast alloys. Brass is made up of 5 to 40 percent of zinc and the rest copper, depending on the use to be made of the brass. Interestingly enough, the same metals also compete with zinc in its many end-use categories.

Substitution possibilities for zinc, as for any other metal, depend on (1) availability of the substitute materials in sufficient quantities, (2) similarity in technical properties, and (3) relative cost of replacement.

In terms of similarity in technical properties, aluminum and magnesium are the only major metals that can be used for coating steel. The relative price of magnesium in the past did not attract any substitution. Moreover, magnesium has not yet been available in sufficient quantities to convince the consumer to replace zinc in its major uses.[18] Aluminum has been used to some extent for coating steel but is technically inferior to zinc in protecting steel from atmospheric corrosion.[19] In die casts as well, technically speaking, aluminum and magnesium are the only promising substitute metals for zinc. Given an adequate increase in the relative price of zinc vis-à-vis these metals, aluminum, which is available in sufficient quantities, may in the long run prove to be a suitable substitute for zinc on a larger scale. However, relatively easy machining and greater precision obtained with zinc will limit the substitution of aluminum to some extent.

The use of zinc for brass and rolled zinc, as noted above, is more popular for construction purposes in the European countries. This is mainly because of traditional preferences which may change with time, or with a sufficiently large change in relative prices. Use of brass for munitions depends more on political or military factors, and on technological developments in the armaments industry. The use of rolled zinc in lithography does not attract any substitution, as zinc is the most technically suitable metal for the purpose. Similarly, use of zinc for zinc oxide in the rubber and paint industries, which use three-fourths of the available zinc oxide, hardly attract any substitution, again owing to some technological factors.

Thus, the substitution possibilities for zinc are very limited, at least in the short run. In the long run, of course, no material is indispensable if a sufficient change in relative prices warrants such a substitution. However, the cost of zinc in the end product is often so small that even a relatively large change in the price of zinc may take some time to convince zinc consumers to replace it with some other material. This is corroborated by estimates of demand price elasticities—to be discussed in detail in chapter 5—which are very low. These estimates, though statistically significant in the long run, are close to zero or often wrongly signed in the short run. In particular, the uses of zinc for die casts, brass, and the miscellaneous categories, as would be expected from the above discussion, show rightly signed high elasticity values in the long run, though the short-run elasticity values for these end-use categories are either zero or very low. This pattern of consumption response may be attributed to the facts that (1) changes in relative prices of zinc have not crossed the permissible range for nonsubstitution,[20] and (2) technological factors predominate for the consumption of zinc in many areas.

Future Developments

The consumption of zinc in the future may be adversely affected by factors such as a reduction in the relative prices of the existing substitutes, and some technological or institutional developments. A few of such developments, which may be speculated at present, may be noted as follows:

1. a development of alternative materials such as hard plastics at relatively attractive prices,

2. a change in the nature of the end product itself,
3. a development of more suitable alloys,
4. a change in the traditional use of zinc in some final products.

Hard plastics which have already been developed, such as ABS, polyamide (nylon) 6/6, polycarbonate, polypropylene and polystyrene, may be used for die casts where a longer life of the die is not a binding requirement. In fact, French automobile producers were already using some of these plastic materials in the late sixties for radiator grills, air vents, and so forth, and were expected to extend the use of these plastics to some other automobile parts. The recent price rise in zinc and other metals further encouraged the consumers in Japan and some other countries to think seriously about substituting plastics for zinc. However, the accompanying rise in petroleum prices has inhibited such a substitution. Besides their use for die casts, plastics can also substitute for zinc in some of its constructional applications such as coating steel, piping, and others.

A change in the nature of the final product, such as a shift in preference towards smaller and/or light-weight cars induced by higher oil prices, or introduction of electric cars or municipal mass transit systems induced by environmental considerations, would adversely affect the growth in zinc consumption. Further, technological developments in producing special kinds of steel immune to atmospheric corrosion, or some other metal alloys which could possibly replace zinc in some of its major uses might reduce consumption of zinc in the future. On the other hand, technological necessity and/or the superiority of zinc over other metals in many final products, as well as individual tastes and preferences for zinc over other materials such as plastics, could, in the future, restrict the possible reduction in growth of zinc consumption. Technological developments in the metallurgy of zinc may make it more desirable in some uses. Research and development as carried out by some research organizations such as the International Lead and Zinc Research Organization may expand the consumption of zinc by finding new applications.

A priori, therefore, it is difficult to predict which factors will be stronger in influencing the consumption of zinc in the future. However, it is clearer that the future growth in the consumption of zinc may not be very significant provided that zinc prices in the future remain relatively stable, and technological development in zinc does not lag behind development in other competitive metals and materials.

International Trade

As may be expected from the previous discussion of demand and supply aspects, the international market for zinc may be divided into two parts, one for zinc ore and the other for zinc metal. Some countries have excess production of zinc ore in relation to their available smelting capacity, and this brings them onto the world market to sell their ore to countries with excess smelting capacity relative to mine production of zinc. The smelting capacity in relation to the consumption requirements of the metal in a country determines its position (buyer or seller) in the world market for zinc metal (see table 2-5). This is based on the presumption that the cost of transporting ore and metal from one country to another will make it necessary to satisfy the maximum possible requirements from within the country. Besides, consumers of zinc may prefer to obtain their metal requirements from the local producers for the following reasons: (1) they may feel more secure by obtaining the metal in the local market in the event of emergencies such as war or general shortages, or an unforeseen increase in the demand for their

Table 2–5
Supply-Demand Imbalances in the FME World, 1960 and 1974
(thousand metric tons, zinc content)

Country/Area	*Mine Production (1)*	*Metal Production (2)*	*Metal Consumption (3)*	*Mine Balance (1) − (2)*	*Metal Balance (2) − (3)*
1960					
Australia	277	122	93	+155	+29
Canada	367	237	51	+130	+186
Mexico	271	53	30	+218	+23
Peru	167	32	2	+135	+30
United States	408	792	790	−384	+2
Japan	148	181	189	−33	−8
Europe	540	123	1,127	−383	−204
Rest of world	232	99	172	+133	−73
Total FME world[a]	2,410	2,439	2,454	−29	−15
1974					
Australia	386	284	120	+102	+164
Canada	1,113	438	137	+675	+301
Mexico	263	133	60	+130	+73
Peru	357	71	17	+286	+54
United States	448	575	1,220	−127	−645
Japan	216	850	678	−634	+172
Europe	736	1,709	1,769	−973	−60
Rest of world	473	302	569	+171	−121
Total FME world[a]	3,994	4,362	4,570	−368	−208

Source: "Zinc," *Mineral Bulletin*, MR 159 (Ottawa: Department of Energy, Mines and Resources, 1976), p. 31. Reprinted with permission.

[a]Balances for total FME world are accounted for by scrap supply, stock changes, and trade with CPE world.

products,[21] (2) they may have business interests in the locally produced metal,[22] or (3) their specifications may be more adequately met by local producers.

World Market for Ore

In 1960, Australia, Canada, Mexico, and Peru together accounted for about 83 percent of the total free world exports of the mine production of zinc. Although the market shares of the individual countries changed over the 1960–1974 period, the total share of these four countries in the FME world exports remained constant. Canada and Peru increased their shares from an equal figure of about 17 percent to about 46 and 18 percent, respectively. Australia and Mexico stood as the main losers as their shares declined from 20 and 28 percent to 12 and 7 percent. Canada, thus, occupies a leading position with 46 percent of the total free-world sales of zinc ore in the world market (see table 2–6). At the same time, it may be noted that total exports as a proportion of total mine production remained fairly constant at approximately 35 percent.

The United States and some more industrially developed European countries were the main buyers of zinc ore, absorbing more than 90 percent of the total zinc ore marketed in 1960. Both the United States and the European countries as a whole shared equally in this market for zinc ore. By 1974, whereas the share of the European countries as a whole remained stable at 47 percent, the U.S. share declined considerably to about 13 percent only (see table 2–6). The primary reason for the decline in the U.S.

Table 2–6
FME World Zinc in Concentrate Trade, 1974
(thousand metric tons)

	Importers					
Exporters	*Europe*	*Japan*	*United States*	*Others*	*Total*	*Percentage of Total*
Canada	411	196	164	51	822	46.4
Australia	93	114	—	5	212	12.0
Peru	156	135	13	14	318	18.0
Others	179[a]	122	52	66	419	23.6
Total	839	567	229	136	1771	—
Percentage of total	47.4	32.0	13.0	7.6	—	100.0

Source: "Zinc," *Mineral Bulletin*, MR 159 (Ottawa: Department of Energy, Mines, and Resources, 1976), p. 35, and *Lead and Zinc Statistics* (United Nations: International Lead and Zinc Study Group, June 1975).

[a]Excluding inter-European trade (407,000 tons)

share has been the closure of many smelters in the late sixties and the early seventies.

World Market for Zinc Metal

The United States alone absorbs about one-half of the metal sold on the world market. This is in contrast to about 2,000 metric tons that the United States exported in 1960. Again, the main reason is the recent closure of many smelters in the country. Imports in both Japan and Europe declined substantially over this period. The European countries that together absorbed more than one-half of the world zinc metal from the world market (excluding intra-European trade) have observed a decline in their market share to only 15 percent. Japan, on the other hand, has not only become self-sufficient in its metal requirements, but also now supplies a substantial proportion (18 percent in 1974) of the total metal marketed in the FME world (excluding intra-European trade). See table 2-7.

Besides Japan, the major suppliers of metal in the world market are Australia, Canada, Mexico, and Peru, constituting approximately 60 percent of the metal marketed (excluding intra-European trade). This is in contrast to their share of about 85 percent in 1960. The decline in their share has mainly been gained by Japanese smelters.

Thus, about one-fourth of the total zinc metal and one-third of the total zinc ore enter the world markets. On the seller's side, in both the zinc ore and

Table 2–7
FME World Zinc Metal Trade, 1974
(thousand metric tons)

	Importers				
Exporters	*Europe*	*United States*	*Others*	*Total*	*Percentage of Total*
Canada	32	239	25	296	29.0
Australia	14	38	109	161	15.7
Peru	8	28	30	66	6.5
Mexico	23	35	16	74	7.2
Others	79[a]	169	178	426[b]	41.6
Total	156	509	358	1023	—
Percentage of total	15.2	49.8	35	100.0	100.0

Source: "Zinc," *Mineral Bulletin*, MR 159 (Ottawa: Department of Energy, Mines, and Resources, 1976), p. 35, and *Lead and Zinc Statistics* (United Nations: International Lead and Zinc Study Group, June 1975).

[a]Excluding inter-European trade (398,000 tons)

[b]Japan = 172,000 Tons (18%)

zinc metal markets, Australia, Canada, Mexico, and Peru, with Canada playing the leading role, are the main suppliers of about 80 and 60 percent of the total ore and metal, respectively, entering the world trade. On the buyer's side as well, the zinc ore market is fairly concentrated, with some European, Japanese and U.S. smelters accounting for about 90 percent of the total ore traded. This concentration on the ore imports side in the future is, however, likely to decrease as the mining countries increasingly smelt their own mine product. We shall discuss this aspect in detail in the next chapter.

The Price System

Zinc is an internationally traded homogeneous commodity.[23] The international prices, therefore, in the different countries should be proportional to differences in the costs of transportation and tariffs, if any. However, the price system in the zinc industry reflects various institutional characteristics and, therefore, requires more careful study. We shall, for convenience, divide the discussion of the price system into price structure and price behavior.[24]

Price Structure

At present, there is a three-tier price system in the international market for zinc:

1. The London Metal Exchange (LME) price
2. Producer basis price that prevails outside the United States (usually referred to as the Commonwealth producers' (CWP) price
3. The U.S. producers price (USP)

The London Metal Exchange, which was founded in 1882, has since then developed and maintained the position of a terminal market for all the major nonferrous metals except aluminum. Each day four prices are issued for all these metals—the buyers' and sellers' prices for delivery the day following, and for delivery in three months. The coexistence of the LME with long-term contracts in ore and metal provides an opportunity for the consumers, who are long or short in terms of quantities, to correct their position. The forward market at the LME allows the buyers and sellers to hedge against the short-term price fluctuations. Thus, a proper functioning of the LME can serve both as a sensitive indicator of supply and demand imbalances and as an instrument for stabilizing short-term market fluctuations. However, the market has exhibited very unstable behavior over time, especially with the erosion of prices in 1957 and their escalation in 1964. This led to the establishment of a fixed-price system by the major ore and metal producers outside the United States towards the end of 1964. Since then, the turnover on the LME has fallen substantially, now accounting for only 10 percent of the total FME world trade in recent years.

After a relatively long period of stable prices, the price of zinc at the LME doubled from an average of £76.8/ton in 1963 to £155/ton in late 1964. This caused great concern among consumers, whose requirements were drawn on the earlier prices, and fear among producers of substitution by other metals such as aluminum and steel, which are known for relatively stable prices. The lack of adequate stocks to control such a price hike among the metal producers led the major non-U.S. producers to agree upon a fixed-price system. In fact, the Imperial Smelting Company of the United Kingdom was the initiator of this move and was supported by the important producers in Australia and Canada. Later, the other smelters of the European continent also agreed to join these major Commonwealth producers. Initially, on 13 July 1964, this combine of the Commonwealth producers fixed the price at £125/ton. Fear of substitution even at this price led them to reduce the price to £110/ton within two months. Since then, both the LME and the CWP prices have moved together, until recently. The LME price, of course, as it is very sensitive to demand-supply imbalances, had a larger fluctuation. After about ten years of relative stability, the LME price has again shown a very sharp upswing, by about 400 percent within the one year 1973–1974. The continuing high demand and the closure of many smelters in the United States also encouraged the Commonwealth producers to raise their price by about 100 percent during the year. By the end of 1974, the LME price again dropped in line with the CWP price and once again both prices started moving together. Although the turnover on the LME, as noted, has decreased to about 10 percent of the total trade, it still remains the price which is market-determined and thus is a better indicator of the market situation.

The CWP price continued to be quoted in pounds sterling for the Good Ordinary Brand (98 percent purity—G.O.B.) zinc metal until 1975–1976, when the severe erosion of the pound sterling promoted a switch to the more stable U.S. currency. All the producers of concentrate and primary metal outsde North America base their sales on this price, which is quoted as CIF world port basis. Delivery of the metal to inland customers is based on this price, in the domestic currency equivalent, plus any additional costs of freight, duty, and grade premiums (more than 98 percent purity). These additional costs to the customers are based usually on negotiations between individual sellers and buyers and the concurrent market conditions. *Metal Bulletin* (London, England), reports the producers' price every day. It changes only when price changes have been announced by the major metal producers in Australia, Canada, and Europe having a combined smelting capacity of more than 1 million tons per year.

The U.S. producers' price for "prime western" grade zinc (98 percent purity), is published in *Metals Week*. This is based on the prices announced by the U.S. primary zinc producers. The price is a weighted average that reflects both the prices charged and the sales made by the individual producers. This same metals price also forms a basis for the price of concentrates sold to smelters in the United States. Since 1972 the foreign producers have set up a separate price for the sale of zinc in the United States which is often competitive to the U.S. producers' price. The U.S. producers' price has also often been influenced by the various U.S. government policies such as tariffs, quotas, the stockpile program, and some additional incentives to smaller companies. These policies will be discussed in the next section.

Although the price of concentrates to a large extent depends on the price of metal itself, the pricing system for concentrates is a little more complicated. The market price of the concentrate depends on (1) the actual price of the metal, (2) costs of the recovery of the metal and the associated byproducts and coproducts (which in turn depend on the nature of ores and the smelting process employed), and (3) the benefits of the associated metals to the smelters. Smelters usually pay for 85 percent of the zinc content in the concentrate, although recently recovery of the metal has substantially improved due to more advanced smelting processes. The smelting charges often represent about 37 percent of the payable zinc value. The payment for other metals recovered in the process depends on the recovery of these metals, cost of recovery, and the market conditions. Usually the sale of concentrates is based on two- to three-year contracts with delivery at the buyer's works or the CIF ports of discharge depending on the agreement between the buyers and the sellers.

Price Behavior

Although the London Metal Exchange, the terminal market for most nonferrous metals, was opened in 1882, its free operation was severely hindered during the interwar period because of various political factors. It resumed its free operation only in late 1953. The discussion in this section will be limited to the period after 1955 only, leaving the discussion of the earlier period to a historical appendix to this chapter.

The behavior of prices during this period (plotted in figure 2–1) can be more conveniently analyzed by dividing this period into 1955–1964, when the world zinc market had a two-tier price system, and the 1965–1975 period, which could be characterized by a three-tier price system, as noted above.

1955–1964 Period. If there were free trade between the United States and the rest of the world, one would expect one single price in the world zinc

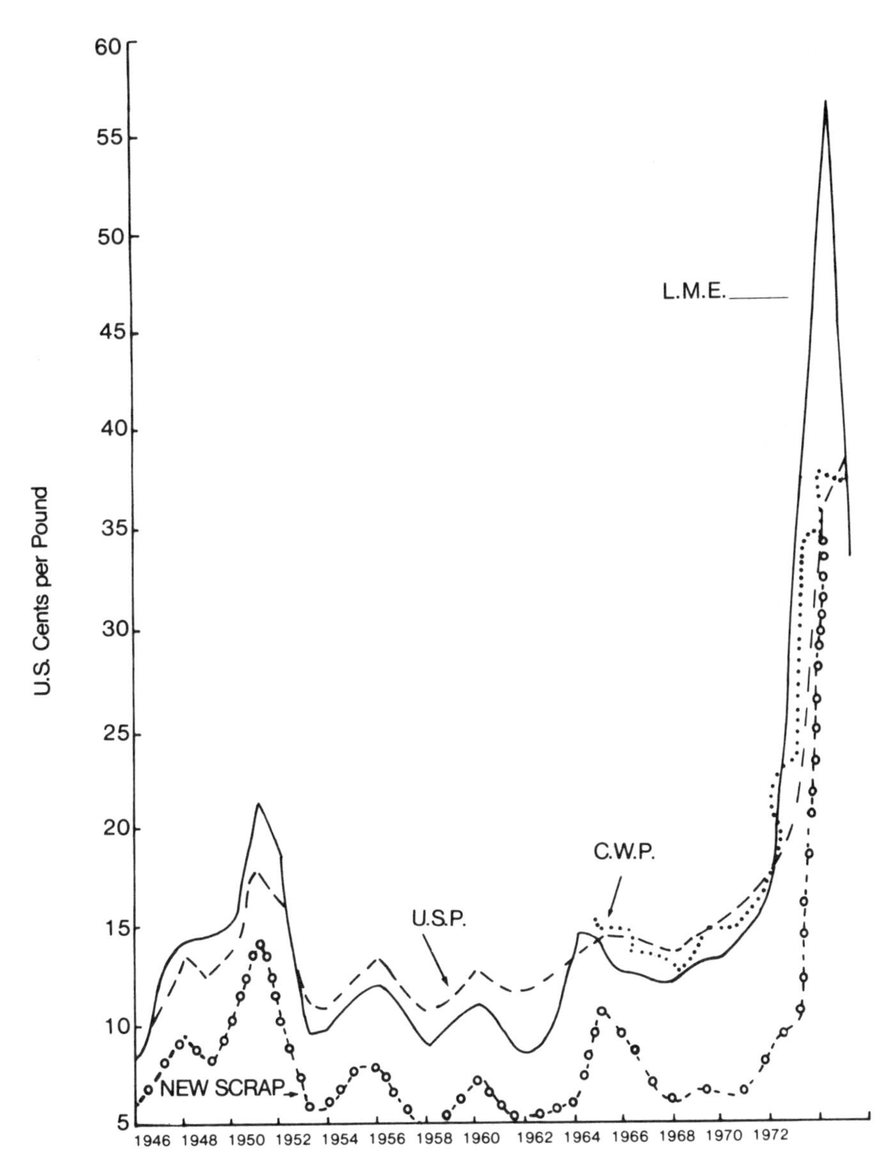

Source: *Metal Statistics* (Frankfurt: Metallgesellschaft Aktiengesellschaft, 1976)

Figure 2–1. Annual Average Prices of Zinc

market varying according to the supply-demand imbalances. The U.S. producer price, however, is influenced by few producers,and supported in its cause by the various U.S. government policies. Nevertheless, as the government support policies have been temporary in nature, the long-run tendencies of both prices have been similar.

Cessation of the high demand of the war period and price controls led initially to an immediate fall in both prices. The surplus of zinc during 1952–1953 was between 70 and 80 thousand tons, leading to a sharp fall in the LME price from 24¢/lb. to 9¢/lb., with a differential of about 2¢/lb. for the U.S. producers' price. With the declaration of purchase of zinc by the U.S. government for its stockpile program up to 300 thousand tons during 1954–1955 and some improvement in economic activity, both prices increased by about 4¢/lb. Producers, in an attempt to take advantage of the recovery in prices, increased their production. This, together with slower growth in activity, expectations of the cessation of the U.S. government's stockpile program, and consumer withholding of purchases in the expectation of a further fall in prices, again created a glut in the zinc market. Both prices fell, but this time the differential between the LME and the USP prices widened from 2¢ to about 3¢/lb. This attracted large imports into the United States at a time when the industry was going through a recession period.

During the same period, the Raw Materials Commission of the UN was considering the feasibility of concluding a raw materials agreement to avoid wide fluctuations in zinc prices. Subsequently, the UN called a lead and zinc conference in London to alleviate the problems through new methods to balance the supply and demand. In 1958 this group recommended a voluntary cut in the mine and smelter output, and this was agreed upon by the major producers. As is apparent, these cuts, rather than being dictated by cost considerations, were dictated by price considerations, and if successful, would have led to doubt as to the competitiveness of the industry.

At the same time, however, the U.S. government decided to impose quotas on imports of zinc. This nullified the attempts of the producers in the rest of the world to stabilize prices at a reasonable level by restricting output. The quotas were removed in late 1965. During the quota period, the LME and the USP price differential rose to more than 4¢/lb., thus providing an opportunity for U.S. producers to realize monopoly profits.

1965–1975. As a consequence of an economic recovery in the United States and to some extent in Europe in the early sixties, the LME price more than doubled during 1963–1964. As noted above, this alarmed the major zinc producers, who foresaw the possibility of substitution of other metals and plastics for zinc, and hence led the Commonwealth producers to introduce a price that would not cross the range of nonsubstitution and would be stable. Later, all other major European producers joined this system.

In the short run, substitution elasticities are very low owing to the technological nature of zinc and the industries using this metal. Plants built for using a particular metal such as zinc may not be used for other metals without substantial modification which may be both costly and involve a relatively long time lag. In the long run, however, when technological aspects of the plants can be adjusted and a near substitute found satisfactory, the zinc industry can incur significant losses by losing markets. This long-run profit-maximization behavior with stable prices seems to be the major objective of the zinc producers and the main reason for deviation from the simple free-market mechanism of the LME. The success or failure of this attempt then depends on their organizational structure—a subject to be taken up in the next chapter for detailed investigation.

In the past a similar attempt was made in the copper industry. The fear of substitution due to high prices of copper at that time led the Roan Selection Trust (RST) single-handedly to offer copper for sale at lower prices. However, RST at that time failed to persuade the other producers to sell copper at the RST-announced prices, and hence had to abandon her attempt within two years. The RST, with the cooperation of the Anglo-American Consolidated, made another attempt in 1964. However, noncooperation by Chilean and African producers led to the same fate within two years.[25] Zinc producers, on the other hand, have been quite successful in their attempt, as shown by our experience during the 1963–1975 period. Initial cooperation of all the major producers has continued in cases of both lowering and raising zinc prices. Production cuts and/or stock changes were undertaken voluntarily many times to keep the prices higher in recessionary situations.

There have been substantial, though temporary, differences between the CWP and the LME prices at times. This, however, does not seem to have worried the major producers very much as turnover on the LME market is very small, major sales are done on two- to three-year contract bases, and the short-run elasticity of demand is very low. The CWP price that was reduced to £110/long ton in September 1964 remained unaltered for about one and a half years. The LME price during this period gradually declined and became in line with the CWP price. The CWP price was further reduced to £102/long ton in April 1966 in response to recessionary forces and continued at that level for about a year. The LME price during this year fluctuated in a narrow band of about £4/long ton. Cessation of the Vietnam War in the late sixties and the prevalent recessionary forces induced both production cuts and a lower CWP price at £98/long ton (equivalent to £114/long ton after the devaluation of the British pound in 1967). This price was maintained for about one and a half years during which the LME price fluctuated below the CWP price up to about £4/long ton.

The U.S. producers, however, increased their price by about 1¢/lb. in late 1964 because they were protected by the U.S. quotas. Somehow, substitution possibilities did not worry the U.S. producers, as they thought they could maintain a stable price. The economic conditions in early 1967 that led the European producers to reduce their price, induced the U.S. producers (after a record stability of two and a half years at 14.5¢/lb.) to decrease their price to 13.5¢/lb.

In early 1969 the CWP price was increased to £130 in a long-overdue response to improved economic conditions. It was maintained at that level throughout the recessionary period until the middle of 1971, when it was again increased to £152 to cover rising production costs. The LME price during this period kept fluctuating according to the economic conditions. The U.S. producers' price followed the increases in the CWP price. Thus, it is apparent that cooperation among the CWP producers on the one hand and the U.S. producers on the other, helped to maintain their prices relatively stable compared to the unstable free market (LME) price.

During 1973–1974, both the CWP producers and the U.S. producers raised their prices by more than 225 percent. Whereas the LME prices during this period, as in 1963–1964, fluctuated widely (as low as £164/metric ton in January 1973 to as high as £738/metric ton in May 1974), the CWP and the USP prices, rather than fluctuating up and down at short intervals, increased by big jumps together. By the end of 1974 the LME price again dropped in line with the producers' determined prices. The rise in the producers' prices may be attributed to a combination of many factors such as (1) the rise in costs, (2) the closure of many smelters in the United States, and (3) the short-run profits induced by the business boom in this period.

Whereas the copper industry producers in similar circumstances failed in their attempts to stabilize their price, the zinc industry producers have been successful. This is surprising, as there is no evidence of any formal cartelization in the zinc industry as compared to those in other nonferrous metal industries where formal cartelizations were introduced at one time or another. Further, none of the zinc industry specialists connected with policy formulations in the industry believe in any lack of competitive environments within the industry.[26]

As is apparent from this record of the movements in the LME prices, daily and monthly variations in prices, as exhibited on the market, are uncontrolled. Price changes by producers are planned, however. These planned price variations are maintained by manipulating stocks and sometimes production itself. The price changes are undertaken only when the majority of the producers, acting together as monopolists or a combine, agree on the move. These price changes may be induced either to maintain the price of zinc in the nonsubstitutable range, or to take advantage of improved market conditions. Both profits and losses are tolerated in the short run. Long-run profits are maximized. A priori, these results are more likely in an imperfectly competitive market environment. However, whether the market environment has been far away from a perfectly competitive one requires a deeper study of the organizational structure of the industry—a subject of the next chapter. Whether the solution in terms of prices and output arrived at by the producers, that is, the market behavior of the producers, is much away from a perfectly competitive one is an empirical question to be taken up in the later chapters.[27] Meanwhile, however, it is possible to look more closely at the national and international policies which may have been responsible for the existing price and output configuration.

National Policies and the International Market[28]

As noted, the world zinc market has been divided into the U.S. zinc market and that of the rest of the world. The separation of the U.S. zinc market is due to the existence of monopolistic elements (a few producers in the United States) and their protection by various U.S. government policies over time. In the rest of the world, some national governments and international organizations have considered the possibility of applying policies such as buffer stocks and tariffs to manipulate the market in their favor also. The national policies in the rest of the world have been rather insignificant, however, both in relation to the world market and to U.S. government policies, and hence will be mentioned only briefly. Some international policies that have led or are likely to lead to distortions in the market will be reviewed in the next chapter.

The European Common Market countries have several times considered and actually levied duties on the import of materials. However, in the case of zinc their success in distorting the market to any significant extent has been arrested because of their very large dependence on foreign-produced zinc ore. France and Japan have recently considered the possibility of instituting a stockpile program similar to that of the U.S. government. France, in April 1975, signed a decree to establish a security stockpile for imported metals and ores to be used in crisis periods such as blockade by war. This is a ten-year program estimated at Fr. 1 billion and includes all strategic metals, thus leaving the share of zinc as not very significant for the world market. Further, although nearly self-sufficient in zinc metal, France is almost totally dependent upon foreign concentrates for her refinery production. Similarly, the Japanese government proposed a ¥30 billion program for stockpiling refined copper, lead, zinc, and aluminum in the 1976 fiscal year. However, this program at present appears to be only a temporary measure to help the producers of these metals, who were holding large inventories.

The U.S. government has tried to help the domestic producers of zinc through various measures.

> The U.S. government, by various congressional actions and administrative programs, has regulated war time production and consumption of zinc, purchased and sold zinc under the strategic and critical materials for

stockpiling act, subsidized exploration and production from small mines, and imposed limitations on imports.[29]

The U.S. Government Stockpile Program (USGSP)

The stockpile program of the U.S. government has been instituted under a triple scheme consisting of the Strategic and Critical Materials Stockpiling Act (1946), the Defense Production Act (1951), and the Barter Program of the Commodity Credit Corporation. The stockpile program, although sometimes helpful to the zinc industry in its pursuit of growth, has often been reported to be a destabilizing factor in the world zinc market.[30] Frequent changes in the stockpile programs and ambiguity in the objectives of the U.S. government are considered to be the main source of instability.

During the period 1946–1953, the stockpile policy was mainly concerned with the objective of securing a sufficient stock of minerals to cope with the possibilities of war or hostile behavior by some major producers in the industry. Under the umbrella of a heavy stockpile program in the United States, the U.S. zinc industry in particular, and the world zinc industry in general, grew much more than would be warranted by normal market conditions.

The end of the Korean War boom left the industry in miserable condition at the end of 1953. The price of zinc metal in the United States declined from 21¢/lb. (24¢ at the LME) to 11¢/lb. (9¢ at the LME). The rate of acquisition of the metal for the U.S. stockpile was too slow to keep alive many mines that had grown out of proportion because of heavy war requirements and the stockpile program in the earlier period. The U.S. producers through the "Escape Clause" petitioned for higher tariffs. The U.S. government, on investigation, granted an accelerated stockpile program instead of the demand for higher tariffs. This kind of change in objective behind the stockpile program was severely criticized by various groups in the United States and abroad. An editorial comment in the *Engineering and Mining Journal* (October 1954, p. 71), though humorous, is remarkable in showing the feelings towards this government policy: "if the idea is accepted, there is no reason why the government should not stockpile "excess" shoes, television sets, or even automobiles, if it should become necessary to stiffen the domestic markets.... The excess stocks could always be sold at a price in the depressed areas of the world."

Economic recovery in 1955–1956, voluntary cutbacks in production by many non-U.S. producers, and some extra demand from the increased stockpile program helped the zinc industry in its recovery. However, the recovery was only temporary. The U.S. stockpile of zinc had reached 1.50 million short tons by 1958. Any further increases in the U.S. stockpile were

Table 2–8
U.S. Government Stockpile of Zinc, 1945–1975
(Short tons)

Year	*Inventory*[a] *as of December 31*	*Net Purchases/ Releases (−)*	*Declared Objective*
1945	0	0	
1946	69,223	69,223	1,300,000
1947	93,381	24,158	—do—
1948	490,595	397,214	—do—
1949	594,657	104,062	—do—
1950	644,146	49,489	—do—
1951	649,163	5,017	—do—
1952	661,714	12,551	—do—
1953	700,320	38,606	—do—
1954	842,463	124,143	—do—
1955	966,551	142,088	—do—
1956	1,147,710	181,159	—do—
1957	1,462,023	314,313	—do—
1958	1,548,235	86,212	—do—
1959	1,583,564	35,329	—do—
1960	1,578,719	4,845	—do—
1961	1,579,616	879	—do—
1962	1,579,907	291	160,700
1963	1,580,941	1,034	0
1964	1,505,234	(−) 75,707	0
1965	1,312,868	(−)192,366	0
1966	1,212,368	(−)100,500	0
1967	1,198,122	(−) 14,246	0
1968	1,160,606	(−) 37,516	0
1969	1,142,185	(−) 18,421	560,000
1970	1,141,490	(−) 695	560,000
1971	1,137,937	(−) 3,553	560,000
1972	949,583	(−)188,353	43,944[b]
1973	675,589	(−)273,330	37,698
1974	390,780	(−)284,809	19,087
1975	384,905	(−) 5,875	10,861

Source: *The U.S. Zinc Industry: A Historical Perspective* (Washington D.C.: U.S. Bureau of Mines, IC 8629, 1974), p. 74, and *Yearbook* (New York: American Bureau of Metal Statistics, 1970–1975).

[a]DPA inventory not included (data not available)

[b]Computed as (total stockpile minus uncommitted stockpile)

due to the emergence of recessionary tendencies in the United States and some other countres, which together threw the market into the hands of speculators. The U.S. price, which had risen to 14¢/lb. by 1957 (LME 13¢/lb.) steadily declined to 11¢/lb. at the LME (lowest in the post-world-war period) by mid-1958.

Although the stockpiles were not released during this panicky period, the U.S. government stopped its accelerated stockpile program. As a substitute to the stockpile program, for the protection of the U.S. zinc industry the government imposed quotas on imports. The zinc industry in the rest of the world, as a result, was left grumbling only at their folly in having expanded their production in response to the stockpile program. In 1960, out of 1.58 million short tons of zinc accumulated in the stockpile, 1.4 million short tons was officially declared as surplus, subject to be released in "due time;" and the objective of the stockpile in 1963 was reduced to zero (see table 2–8). This wavering attitude of the U.S. government about the stockpile policy created some instability in the international market for zinc. In June 1976, the U.S. administration announced plans to issue a new, and generally higher stockpile objective that would include upgrading much of the metal forms currently held in the stockpile. With upgrading, the government will have to change its posture of the past ten years to become a purchaser of the metal. This could, to some extent, help the smelters of the United States, half of which closed between 1969 and 1973.

The U.S. Tariffs, Quotas, and Other Incentive Programs

Tariffs. The history of tariff duties on zinc in the United States is now more than a century old. The first tariff was enacted in 1846, since when it has been changed many times.[31] From 1951, however, duties on zinc ore and metal have been more or less stationary at $0.67/lb. for the ore, and $0.70/lb. for the metal. In spite of much periodical lobbying by the U.S. producers for increased duties, the U.S. government has opted for what is politically claimed to be the more viable means of protecting the domestic industry, such as the stockpile program, quotas, and some direct incentives to the producers.[32] In 1975 the tariff schedule was amended to allow zinc in ores, scrap, and waste to enter duty free until 30 June 1978. This amendment came in response to a severe shortage of zinc ore, which had led to the closure of many smelters earlier.

Quotas. While the zinc industry was passing through a critical period of low prices coupled with abnormally excessive capacities around 1957, the U.S. producers were petitioning for higher tariffs again. Consequently, by Presidential Order, effective October 1958, import quotas were granted. These quotas were based on an average of 1953–1957 imports. The coverage given by the quotas encompassed all lead and zinc ores, intermediate smelter products, and the metals. Nevertheless, the way in which the quota program was implemented received much criticism both from academic and business circles. First, the quota legislation did not make any provision for the transfer of imports from one country to the quota allocation of another country. Consequently, one finds the coexistence of unfilled allocations and zinc shortages in U.S. zinc market. Second, quotas on both ore and metal implied that the nonintegrated smelters would receive much less advantageous positions as compared to the integrated smelters. Third, quotas did not have any flexibility to accommodate themselves to economic changes in the country. For instance, until 1961 quotas were not effective as they were too high compared to the requirements for imports in such a recession period. In contrast, during the latter part of the quota period (1962–1965; quotas were removed in October 1965), when economic recovery induced more demand for imports, quotas were found to be too restrictive.[33]

However, whatever the limitations of quotas, the U.S. producers did earn high profit margins as shown in the differential between the U.S. producers' price and the price at the LME. On the other hand, the non-U.S. producers were trying hard to find markets for their products and often had to cut back their production levels.

Besides tariffs, quotas, and stockpile programs, the U.S. zinc industry also received governmental protection through various assistance schemes instituted by the U.S. government. These assistance schemes took the form of subsidies to provide stabilization payments to small producers, different depletion allowances on domestic and foreign production, sharing costs of exploration, and other similar other economic incentives.

In addition to its protectionist policies, the U.S. government has also taken certain measures which have adversely affected the U.S. zinc industry. These measures were induced mainly by considerations of environmental protection. This legislation has increased the cost of production to some extent, thus making the U.S. metal industry less competitive. This was the main cause for the closure of many smelters in the late 1960s and early 1970s. However, similar legislation is likely to be enacted in the other countries, making its overall effect less significant for the competitiveness of the U.S. industry. More important may be the fact that the materials which are promising substitutes for zinc and do not require such environmental legislation will have improved their competitive position vis-à-vis zinc.

Notes

1. This section is based on the statistical information collected in Roskill (1974), chapters 2–4 and *Metal Statistics* (1950–1975), unless otherwise indicated.

2. See Behrman (1976) and Eckbo (1975).

3. For example, during the interwar period many mines continued to operate for many years in spite of higher average variable costs as compared to average revenue. See appendix to chapter 3.

4. Both the European and Australian mines are relatively old. Many

deposits have reached exhaustion in both continents. However, Ireland in Europe and a few new mines such as Hilton and Lady Lorretta in Australia seem very promising at present. Opening of these mines has been postponed for various political and economic reasons. The major producers of zinc concentrate in Europe have been West Germany, Italy, Spain, Sweden, and Yugoslavia. Italy's share over the past few decades has gone down considerably. For a detailed discussion on individual mines in different countries, see Roskill (1974, chapter 4).

5. However, these surface mines of old scrap continually increase with the consumption of the metal, and thus provide an alternative source of zinc for the future. Recovery of zinc from this old scrap material in the future will depend on the cost of collection and method of recovery. Technological developments in the methods of recovery may make this a very attractive alternative source of supply. For a detailed discussion on this aspect, see Roskill (1974, chapter 2).

6. See Roskill (1974, chapter 2).

7. Historical details of these aspects are presented in the appendix to the next chapter.

8. For details, see Roskill (1974, chapter 2).

9. For details, see appendix to chapter 3.

10. Many of these plants were too old and renovations to meet the legislative requirements were found too costly to be undertaken.

11. Kreinin and Finger (1976).

12. Department of Energy, Mines and Resources (1976, pp. 50–55).

13. Data on the consumption structure of zinc have been more extensively collected than data on any other aspect of the world zinc industry. There are numerous sources for the data. However, *Metal Statistics*, Metallgesellschaft, A.G. (one of the oldest annual publications), and *Lead and Zinc Statistics*, a relatively recent monthly bulletin published the *International Lead and Zinc Study Group*, United Nations, are the most reliable with regard to the consistency of data. The analysis in this section is based on the data from one of these two sources, unless otherwise indicated.

14. U.S. Bureau of Mines (1975, p. 10).

15. If the consumers of zinc metal could form a combine or persuade their national governments to do so, there is a great likelihood of an oligopsonistic market structure. However, due to (1) a wider distribution of consumers in any one nation-state, (2) lower importance in terms of cost, and (3) the essential nature of zinc in the final consumer goods, it is not likely for zinc consumers to either form a successful cartel on their own initiative or persuade the governments to join hands for this purpose. These limitations make the concentration of metal consumers ineffective for any market imperfections on the buyers' side in the metal market.

16. See table 2A–1 in the appendix to this chapter.

17. Ibid.

18. Consumers would like to see whether the material, which is a potential substitute for zinc, is available in sufficient quantities, as the substitution of one material for another involves changing plant technology for producing the goods which used zinc previously.

19. See appendix to this chapter.

20. In 1964, in fact, when the zinc prices in the free market were observed crossing the nonsubstitution range, the major producers of zinc ore and metal introduced a fixed-price system now popularly called the European producer's prices. For details see the fourth section of this chapter.

21. See Fisher, Cootner and Baily (1972).

22. For instance, many smelters have their own mines; many consumers of the metal such as steel companies and brass manufacturers have business interests in the local smelting plants.

23. Quality differences, if any, due to particular technological specifications of ore and metal are automatically scaled up or down proportionally in the prices and thus have no implications for the price behavior.

24. This section is based largely on information published in the various issues of *Engineering and Mining Journal* and *Metal Statistics*, unless otherwise indicated.

25. *Metal Statistics* (1965, pp. i–vii).

26. The author discussed this matter with some of the zinc industry specialists.

27. It is very likely that the solution arrived at through prudent planning may be nearer to a competitive solution than the one arrived at in the free market, which is grossly influenced by uncertainty and possible false expectations about the future.

28. The discussion of these aspects is based on (1) Department of Energy, Mines and Resources (1974, pp. 41–43) and various issues of *Metal Statistics* (introductory pages) and U.S. Bureau of Mines (chapters on zinc), unless otherwise indicated.

29. Heindl (1970, p. 817).

30. This is confirmed in the editorial reports and articles in the various professional journals such as *Engineering and Mining Journal*. The U.S. government considered a stockpile policy politically more viable than tariffs for protection of the domestic zinc industry. See *Engineering and Mining Journal* (October 1954, p. 71).

31. For a historical account of tariffs, see the historical appendix to the next chapter.

32. This is the feeling reflected in the various 1962 issues of *Engineering and Mining Journal* during 1958–1962. For example see the issue of February 1959, p. 107.

33. For a detailed examination of the effectiveness of quotas, see Andrews (1970).

Appendix 2A: Technological Aspects of Production and Consumption, and Major Uses of Zinc

Zinc ranks third in consumption among the nonferrous metals of the world, behind only copper and aluminum. It is used as an intermediate good in a wide variety of applications, as shown in table 2A–1. In fact, it is hard to find oneself in the home, in the office, in the factory, or on the street without encountering an application of zinc in one form or the other. Its uses range from construction and transportation to cosmetics and nutrition (see table 2A–2). However, in many instances both the quantity and cost factor in most end products are too small for an immediate recognition of the significance of zinc.

Zinc is often used as a complementary material with the other ferrous and nonferrous metals. Substitution of other metals for zinc depends largely on technical suitability, availability of substitutes in sufficient quantities, and cost considerations. Aluminum, magnesium, and plastics have replaced zinc to some extent in some of its major applications, but substitution of other materials for zinc on a larger scale does not seem to be promising because of

Table 2A–1
FME World Consumption of Zinc Metal, by Sector of Intermediate Demand, 1956 and 1974
(percent)

Sector	*United States*	*Japan*	*United Kingdom*	*West Germany*	*FME World*
Galvanizing	38.2	55.0	26.1	36.7	39.0
	(43.5)	(61.2)	(32.9)	(36.4)	
Die castings	33.2	20.5	19.7	20.4	22.0
	(35.7)	(9.2)	(12.4)	(9.4)	
Brass	13.7	11.0	28.5	23.7	17.0
	(12.3)	(17.5)	(32.0)	(19.7)	
Rolled zinc	3.0	3.7	6.3	14.8	6.0
	(4.7)	(4.9)	(7.3)	(29.5)	
Zinc oxide	5.1	2.8	10.9	1.2	9.0
	(2.0)	(5.7)	(8.4)	(1.8)	
Miscellaneous	6.6	7.0	8.5	3.2	7.0
	(1.8)	(1.5)	(7.0)	(3.2)	
Total	100.0	100.0	100.0	100.0	100.0

Source: *Metal Statistics* (Frankfurt, Metallgesellschaft Aktiengesellschaft, 1957), and *Lead and Zinc Statistics* (United Nations: International Lead and Zinc Study Group, March 1975).

Table 2A–2
U.S. Consumption of Zinc, by the Sector of Final Demand; 1964, 1969, and 1973
(thousand short tons zinc content)

Sector	*1964*		*1969*		*1973*	
Metal						
Construction	393	(38)[a]	526	(38)	533	(38)
Transportation	277	(27)	372	(27)	367	(26)
Electrical	138	(13)	186	(13)	180	(13)
Machinery	104	(10)	139	(10)	140	(10)
Other	126	(12)	171	(12)	185	(13)
Total metals	1038	(100)	1394	(100)	1405	(100)
Nonmetal						
Paint	34	(24)	30		37	(15)
Chemicals	3	(2)	43		60	(25)
Rubber products	75	(53)	98		130	(53)
Other	29	(21)	17		17	(7)
Total nonmetal	141	(100)	188		244	(100)
Total metal and nonmetal	1179		1582		1649	

Source: V.A. Cammarota, H.R. Babitzke, and J.M. Hague, "Zinc, Mineral Facts and Problems," Bulletin 667 (Washington, D. C.: U.S. Bureau of Mines, 1975) p. 10. Reprinted with permission.
[a]Percentage

technical requirements, individual preferences, and the rising cost of other substitute materials along with zinc.

The main threats to the zinc industry of the future exist in the following:

1. technological developments in the plastics industry and in the metallurgy of steel,
2. increased preferences for small cars,
3. development of electric cars or the more efficient municipal mass transit systems induced by pollution controls,
4. rising cost of zinc metal because of environmental legislation, and
5. decreasing reserves of zinc ore over time.

Technical Properties of Zinc[1]

Zinc, a blue-to-grey metallic element, is widely found throughout the world. When freshly cast, zinc has a white-silver-blue appearance, sometimes having been known as "false silver," and on exposure to air forms "an impervious, tenacious, and protective" grey oxide film. Its basic characteristics are:

1. relatively low melting point (419°C),
2. good resistance to atmospheric corrosion combined with a high place in the galvanic series of metals,
3. solubility in copper and some other metals, and
4. inherent ductility and maleability.

These characteristics of zinc have been responsible for its use in galvanizing, die casting, brass, wrought iron, and some other zinc base alloys in the construction and various manufacturing industries. Further, the chemical compounds of zinc, such as zinc oxide, zinc chloride, and zinc sulphate, have found a wide variety of uses in the rubber, paint, and ceramic industries.

Intermediate and Final Uses of Zinc

Galvanizing

One of the largest uses of zinc consists in its provision of protective coatings for iron and steel products. Zinc protects iron and steel from corrosion by a sacrificial action. That is, whenever there are any pinholes or scratches in coated steel, zinc overcoats the pinholes, sacrificing itself to protect the steel from corrosion.

Galvanizing is done by various methods including hot dip galvanizing (immersion of iron and steel products in molten zinc), electrodeposition, metalizing (spraying with droplets of molten zinc), and sheradizing (diffusion of zinc powder into steel surfaces at elevated temperatures). Hot dip galvanizing is one of the oldest, most economical, and most widely used method of galvanizing. However, the electrolysis process is gaining in importance because of the uniformity in coating obtained and the possibility of controlling its thickness.

The major iron and steel products galvanized are sheet and strip, tube and pipe, and wire and wire rope. These products are used primarily in construction for roofing, siding, decking to support concrete floors, heating and ventilation ducts, and similar applications. Galvanized products are also increasingly being used for home appliances, office equipment, automobile door panels, and underbody parts.

Die Casting

Die casting is the art of producing accurately finished parts by forcing molten metal into a metallic die or a mold under external pressure. Zinc, because of its properties of corrosion resistance and low melting point, is found very suitable for die-casts purposes. Comparatively low cost, ease of machining and finishing, and excellent stability have also been important in its increasing use in this area. Today more than one-third of the zinc consumed in the United States is used by the zinc die casts industry.

Zinc die casts are used as trim pieces, grills, door and window handles, carburetors, pumps, door locks, and other mechanical components in automobiles. Such uses consume about two-thirds of all the die casts produced. The other 15 to 20 percent of die casts are used in home appliances, with the rest finding applications in a wide variety of uses such as commercial machines and tools, builders' hardware, plumbing and heating, business machines, office equipment, optical and photographic instruments, timing devices, electronic equipment, and so on.

Brass

The manufacture of brass constitutes the third major area of zinc consumption. Brass is an alloy of zinc and copper with the zinc ranging from 5 to 50 percent, depending on the application of brass. Zinc, when alloyed with copper, combines good physical, electrical, thermal, machining, and corrosion-resistance properties. "Alpha-brasses," that contain up to 40 percent zinc are used for decorative purposes, electrical appliances, cartridge cases, doors, and furniture. Alpha and "beta-brasses" are used for shipping, construction, electrical appliances, and home goods. German silvers, an important group of copper-base wrought alloys containing zinc as an essential element, are used as base alloys for silver-plated flat or hollow tableware and for many other items such as rivets, screws, zippers, optical goods, and costume jewellery. Some casting alloys (alloys of copper, zinc, and one other metal) such as bronzes and various other types of brasses are used for hardware fittings, valves, plumbing fixtures, dairy and soda fountains, trim, ornamental castings, and other uses.

Rolled Zinc

Rolled zinc is produced as sheet, strip, plate, rod, and wire in numerous compositions and alloys depending on the ultimate requirement of the rolled product. Usually a high grade (special high grade, 99.9 percent purity) of zinc is used with copper, magnesium, manganese, aluminum, chromium, and/or titanium as the alloying metals for the alloyed rolled zinc products.

Rolled zinc with a higher purity in the sheet and strip form is used in battery cans, mason jars, eyelets, flashing light reflectors, grommets, cosmetic cases, valleys, facia strips, gravel stops, gutters, organ pipes, casket shells, and so forth. Rolled zinc, with a lower purity, is used for sides and bottoms of dry batteries, roof coverings, cable hangers, counter pois strip, and weather strip. Another important use of zinc is in lithography. Constructional applications of rolled zinc are more popular in Europe.

Zinc Compounds and Zinc Dust

Zinc oxide is the most important zinc chemical with respect to both tonnage and value. In fact, it can be the starting point for all zinc chemicals. Other major zinc chemicals are zinc chloride, lithopone, and zinc sulphate.

Major uses of zinc oxide are in the manufacture of rubber, in paints, and in the ceramic industry. Over one half of the zinc oxide is consumed by the rubber industry. Tires are toughened by a high loading of zinc oxide (about 5 percent in weight) which not only improves the rubber composition's tensile strength and resistance the abrasion, but also protects the rubber by its opaqueness to ultraviolet light and by its high thermal conductivity. Other major uses of zinc oxide are protective and decorative coatings (usually in the form of paints), in photocopy paper, in ceramic products, in cosmetics, in coated fabrics and textiles, in floor coverings, in lubricants, and in many other applications such as agricultural and pharmaceutical products. Other zinc chemicals have similar applications.

Zinc dust, a by product in the distillation of zinc ore, finds its applications in the manufacture of chemicals consumed in the process of printing and dyeing textiles (as reducing agents), in explosives and matches, in tear gas compositions, in the purification of sugar, in treatment of paper surfaces, as a catalyst, as a condensing agent, and in other applications.

Miscellaneous Uses

Approximately 5 to 10 percent of zinc consumed as metal is distributed among a number of miscellaneous uses. Some major uses in this category are sacrificial anodes used to protect ship hulls, submerged steel works and pipes, and as a trace element in animal and plant nutrition.

Substitutes, Complementary Materials, and Future Uses of Zinc

In its major applications, zinc is often used in a complementary manner with other ferrous and nonferrous metals. In galvanizing, for instance, zinc is used for coating steel. Alloys based on zinc invariably use one or more other nonferrous metals. In brass, copper is the main complementary metal used with zinc.

Substitution for zinc depends not only on relative prices of materials but also, and more importantly, on the technical suitability as well as availability of substitute materials in sufficient quantities.

Galvanizing

There is no satisfactory substitute for zinc in this major use. Technically speaking, three metals—magnesium, cadmium, and aluminum—qualify as substitutes. However, cadmium and magnesium are neither sufficient in quantity to take over this use of zinc nor very attractive on cost considerations. Aluminum coatings, although competitive for sheet and strip steel, are termed inferior to zinc coatings. The formation of an insulated oxide film on the aluminum-coated steel is more noble than aluminum itself, which restricts the electrochemical protection of bare iron and steel at cracks and flaws in the coatings. Further, aluminum coatings at present are higher in cost.

Recently, plastic coatings seem to have captured a small portion of market for zinc in this use. However, with rising oil prices and the individual preferences for zinc coatings, plastic coatings do not seem to be very promising substitutes for the near future. The other competitive material for galzanized sheet is aluminum sheet for roofing and siding, and a possible development of low-cost corrosion-resistant steels.

Die Castings

The major substitutes for zinc in die casting are also magnesium, aluminum, and plastics. Uses of zinc base alloys represent about 60 percent of die cast production, with aluminum occupying about one-third of the total die-casting territory. The rest of the die cast production is distributed among various other metals such as magnesium, copper, tin, and lead-base alloys. Zinc-base alloys used for die casting also contain aluminum (3 to 4 percent) and other metals in minor quantities. Thus, the same metals are also complementary to zinc to some extent.

In fact, aluminum and magnesium are important substitutes for zinc in die casting where weight limitations or weight reductions or finishes are important factors. Plastics also have made some inroads into the die-casting territory. However, the use of plastics in die casts are limited to the cases where efficiency and long life of the die is not a binding requirement. Besides, in the future more production of smaller cars, the development of electric cars, or the development of more efficient mass transit systems induced by pollution controls and higher oil prices may decrease the consumption of zinc in this use.

Brass

Copper is complementary in the use of zinc for brass. In the event of higher prices of copper and zinc, a wide variety of materials can substitute for brass in many of its applications. In fact, a large tonnage of brass used in building and marine hardware, plumbing goods, and bearings have been replaced by aluminum, stainless steels, and plastics. However, there are many other uses of brass which depend on public tastes and preferences and hence are difficult to replace with any other material until the relative price of brass is too high.

Rolled Zinc

There are not many suitable substitutes in the applications where rolled zinc is used. In constructional applications, for instance, copper and copper alloys are used for exterior finishings, whereas lead sheets are used for the insulation of sounds and vibrations. Rolled zinc is used mainly for indoor applications such as lining cupboards, covering tables and bench tops, and so forth. In the case of electrical applications such as storage batteries, dry batteries with zinc serving as electrodes are very popular. There are, however, some other batteries that use zinc, cadmium, lead, magnesium, copper, and silver with a wide variety of acid or alkaline electrodes, but none of them is, at present, of any great commercial importance. Further, in these cases other materials act both as substitutes and complements. There is, however, a possibility of further development of rechargeable batteries that may reduce the overall consumption of zinc in this use. The most preferred use of rolled zinc in lithographic plates (for offset press) hardly attracts any substitution of other metals for zinc.

Zinc Compounds

The uses of zinc oxide at present do not attract any substitution by other materials. The rubber industry, the largest user of zinc oxide, depends on the use of zinc oxide for technical properties not easily found in other substitutable materials. Only if synthetic plastics can be developed in the future to make such items as tires with longer life, or if a mass transit system is developed to control pollution, will overall consumption of zinc oxide in this area decrease. Zinc oxide in the paint industry may be replaced by other competitive pigments, but its use may increase if the technology of using it in water-base paints is developed. The use of zinc oxide in the ceramic and cosmetic industries and in plant and animal nutrition can hardly be eliminated by use of any other materials in the foreseeable future.

Miscellaneous

Most other minor uses of zinc fall into this category. Some of these are price-inelastic, whereas others are highly price-elastic. In general, it is expected that these minor uses be more price-elastic than other categories.

Technology and Cost of Mining and Smelting[2]

Although numerous minerals are known to contain zinc, the principal zinc ore mineral is sulphide or sphalerite, popularly known as "blende." These minerals often occur in association with lead (Zn—Pb) or copper (Zn—Cu) or both lead and copper (Zn—Pb—Cu). Except for Canada where Zn—Cu is the predominant mineral form, Zn—Pb is more frequently found in the earth's crust. However, some copper and iron is often associated with Zn—Pb. In addition, most sphalerite minerals have up to 2 percent cadmium and small quantities of germanium, gallium, indium, and thalliam, which are recovered as byproducts at zinc reduction plants (see table 2A–3).

Mining and Milling

Costs and methods of mining differ from one mine to another depending on the nature of deposits, and the stage of the mine. Mixed sulphide deposits in metamorphic rocks, which contain higher percentages of zinc, lead, and copper and are usually found in Canada, require more costly mining methods and separation techniques than the strata-bound deposits in carbonate rocks, which contain a lower percentage of zinc.

Except in the initial stages where open-pit mining can be done, as in some new mines in Canada, the usual method is underground mining. The costs associated with underground mining can be more than 40 percent higher than those for open-pit mining. Further, in underground mining, certain methods such as "cut and fill" are more costly than the other methods. Copper-zinc mines usually require higher-cost methods than lead-zinc mines. The nature of minerals in Canada, Sweden, and Peru require, in general, more expensive methods of mining.

Table 2A–3
Zinc Byproduct and Coproduct Relationship in the United States, 1973

Principal Ore	*Product*	*Unit*	*Quantity*	*Percentage of Total Output*
Zinc	Cadmium	Short tons	3,714	100.0
	Germanium	Pounds	27,000	100.0
	Thallium	Pounds	W[a]	100.0
	Gallium	Pounds	W	W
	Indium	Troy ounces	W	100.0
	Manganese	Short tons	W	W
	LEAD	Short tons	72,000	11.9
	Silver	Troy ounces	3,913,000	10.3
	Gold	Troy ounces	61,000	5.2
	Sulphur	Short tons	267,680	2.2
	Copper	—do—	5,000	0.3
	ZINC	—do—	359,000	74.9
Lead	—do—	—do—	101,000	21.1
Copper	—do—	—do—	13,000	2.7
Fluorine	—do—	—do—	W	W
Silver	—do—	—do—	2,000	0.5
Gold	—do—	—do—	18	Neg.

Source: V.A. Cammarota, H.R. Babitzke, and J.M. Hague, "Zinc, Mineral Facts and Problems," Bulletin 667 (Washington, D.C.: U.S. Bureau of Mines, 1975), p. 13. Reprinted with permission.

[a]W: Withheld to avoid disclosing Company Confidential data.

The winning of zinc after mining starts with milling, which produces zinc concentrates, which in turn are treated at smelter and refinery plants to obtain zinc metal. The milling technique consists of crushing and grinding the mined ore in closed circuits with vibrating and trommel screens and classifiers, which in turn, through differential floatation, separate zinc from gauge minerals. Costs of this process also rise with a greater complexity of the ores.

Smelting and Refining of Zinc

Zinc concentrates thus obtained are sent to smelters either integrated with the mines or which are independent companies set up solely for this purpose. The nonintegrated smelting companies either buy the concentrate from the mining companies or smelt on a toll basis. Quite often, these smelting companies also sell the metal for the mining companies. In 1974, in the FME world, more than 40 percent of zinc concentrates were treated by the independent smelting companies.[3] Inadequate facilities for smelting in the mining companies and the existence of only a few independent smelters in the world (see next chapter) are responsible both for widespread trade in concentrates and probably for some degree of oligopsonistic structure in the market for concentrates.

Smelting technology is much more capital-intensive than that of mining, though the capital requirements for producing one ton of zinc concentrate and smelting the same amount of zinc concentrate are surprisingly similar. The existing smelting technologies in the world can be classified generally into the thermal-reduction process and the electrolysis process. A combination of these two, called the electrothermal process, is also used.

The thermal process, where carbon is used as a reducing agent with the concentrates in horizontal (batch-type process) or vertical (continuous) refractory retorts, produces ordinary-brand zinc (about 98 percent purity). The vertical retort process is more economical and has replaced the horizontal retort process to great extent. This in turn is being gradually replaced by the electrolytic process and the recent blast furnace type, imperial smelting process. The imperial smelting process, though it works on the thermal-reduction principle, has the advantage of being able to treat mixed lead-zinc concentrate and recover both metals, together with any gold and silver present, at very little cost. The electrolytic process, on the other hand, has gained very wide popularity because of its ability to produce very high purity zinc (99.99 percent) with much less environmental pollution. At present, three-fourths of the world zinc smelting capacity is based on the electrolysis process, as opposed to only about 45 percent in 1960. In the future, the electrolysis process, though it will replace the existing old thermal-reduction plants, will, however, require further changes to control the emission of sulpher dioxide, and to meet the environmental laws being instituted in various countries. A consequent rise in the cost of production, in turn, may reduce the competitiveness of zinc vis-à-vis aluminum.

Notes

1. Although there are numerous publications on the technological aspects of metals and minerals containing some discussion of zinc on selected aspects of its technology, the two most important publications with a wider coverage on the technological aspects of zinc are: International Lead and Zinc Study Group (1966) and Mathewson, ed. (1969).

2. For a detailed account of these aspects, see Cairnes and Gilbert (1967), and McMahon et al. (1974, pp. 29–41).

3. However, it may be noted that some of these smelting companies had ownership interests in mining companies in other countries from which they imported the concentrates to be refined.

3 Organizational Structure of the World Zinc Industry

In the last chapter it was observed that the world zinc industry is highly concentrated on both demand and supply sides in terms of the number of nation-states involved. This concentration can influence the world market if the participants in the market are nation-states rather than private producers and consumers, or if the producers and consumers are completely loyal to their national affiliations regardless of what happens to their profit calculations. In the FME world zinc market, it is the private producers and consumers who are the major participants. Furthermore, national loyalties without competitive profits may hardly survive the frequent movements in the world market.

In general then, it may be stated that the organization of a world industry which is most meaningful for market conduct and performance may best be looked into in terms of units of financial control. In fact, the organizational structure in terms of financial control forms the basis of market structure in the current literature on industrial organization.[1] However, state interference either imposed or asked for by the producers themselves must not be neglected. It was observed in the last chapter that interference by the U.S. government exerted great influence on the market behavior of the world zinc industry. Further, it is possible to find numerous examples of international organizations in the zinc industry which involved groups of producers in the countries as units. The European cartels in 1885 and 1928–1934, the American Zinc Institute and American Lead and Zinc Association protecting the interests of American firms producing zinc, and the recent International Lead and Zinc Study Group (an intergovernmental body), all indicate that producers often join hands with national territories as units.[2]

In this chapter focus will rest on the corporate structure of the world zinc industry, in order to provide some guidelines for the market structure and consequently the probable market behavior and performance of the industry. The working of some national and international organizations, insofar as they influence the market behavior, is also discussed at the end of this chapter.[3]

The discussion of the corporate structure of the world zinc industry will be divided into mining and smelting, as both of these differ in their competitive structure, not only because of the countries, but also because of the corporate groups involved. It will be seen how vertical integration, which is increasing gradually and which is also one of the main features of the New International Economic Order professed by the United Nations, can modify our conclusions regarding the structure, behavior, and performance of the world zinc industry as a whole.[4]

Corporate Structure of the World Zinc Mining Industry

According to the list in the 1975 *World Mines Register*, there are about 172 mining sites spread over the world. Eighty mines are controlled by twenty-five companies, and these produce about 85 percent of the world zinc mine output. The majority of these companies are, in turn, located in Canada, the United States, Australia, Peru, and Mexico, in descending order of their shares in the total output. However, the corporate control of these mines gives a very different picture. Eight companies in the world control more than 50 percent of the world zinc mine capacity. Four Canadian companies control one-third of the world zinc mine capacity. A detailed breakdown by countries, however, reveals the situation more adequately. This information is summarized in figure 3–1. (For details of the structure and control of each corporate group involved in the world zinc industry, see tables 3–3 to 3–7 at the end of this chapter).

Canada

Canada possesses about 36 percent of the free-world zinc mine capacity. Three Canadian companies, Noranda Mines Ltd., Cominco Ltd., and Sheritt Gordon Mines Ltd. control about 50 percent of the total Canadian zinc mine capacity. Of the rest, more than half are controlled by the two U.S.-based companies—Texas-Gulf Inc. and Cyprus Mines Ltd. About 5 percent of the total are controlled by an Anglo-American group of S-W. Africa. Thus, we have six companies controlling more than 80 percent of the Canadian zinc mine capacity. On the fringe, however, a number of companies share in 15 percent of the total zinc mine capacity in Canada. Further, Cominco and Noranda together have major interests in one of the largest zinc mine companies of Ireland—Tara Explorations Ltd.—which controls 70 percent of the mine capacity in Ireland. Cominco also has some control or influence in the zinc mining companies in India, Greenland, Spain, and some other countries. The controlling interests of these giant corporations, *not* shown in the table, are increased to a much greater extent through their exploration companies in the third-world countries. Joint interests of Cominco and Noranda in Tara Explorations lead one to doubt their independent behavior in their decision making processes.[5]

United States

Investigation of the corporate structure of the U.S. zinc companies easily refutes the often-cited problem—the nonavailability of sufficient zinc ores in relation to the requirements of the U.S. smelters. In fact, in 1974 control by U.S.-owned companies (or where they had majority interests) of zinc ore production was double that of U.S. smelter capacity, and about 5 percent more than its zinc metal consumption requirements. The three multinational corporations, Asarco, Amax, and Texas-Gulf, controlled about 18 percent of the FME world zinc mine capacity. Of this, 87 percent was located all over the world (outside the United States) including Canada, Australia, Africa, and South America. The total control by U.S.-based companies of world zinc mine production is about 33 percent, a figure comparable to that of Canadian control. The difference in the control structure of the United States and Canada lies, however, in the fact that two-thirds of the U.S. control is outside the country, whereas 85 percent of the Canadian control is of mines located within Canada.

Thus, the North American zinc ore producers control about two-thirds of the FME world zinc mine production. If availability of mineral supplies in the zinc industry were a lever of control over the market, these six companies together could very well influence price and output decisions in the market. However, the degree of vertical integration and their shares in the world trade in zinc may prevent this possibility.

Australia, Mexico, and Peru

Australia and Mexico have one thing in common: their major zinc mines are owned by the outside interests. Three U.S. companies, Fresnillo, Amax, and Asarco, control more than two-thirds of the mine capacity in Mexico, whereas only two companies, Rio Tinto and Asarco, together control more than three-fourths of the Australian zinc mine capacity. Most of the zinc mined by these companies is exported to the U.K. and the U.S. smelters. On the other hand, only 20 percent of the mine capacity in Peru is under the control of foreign companies. The present corporate structure in Peru, however, dates back only to 1968, when, following a military coup, the government issued a mining decree instructing companies holding mining concessions to establish development plans for their concessions or forfeit them to the state. In 1969 Empressa Minera de Peru was incorporated to take over all forfeited concessions and to act as state trading company for all Peruvian mineral products. Before 1973 Cerro de Pasco was the major mine

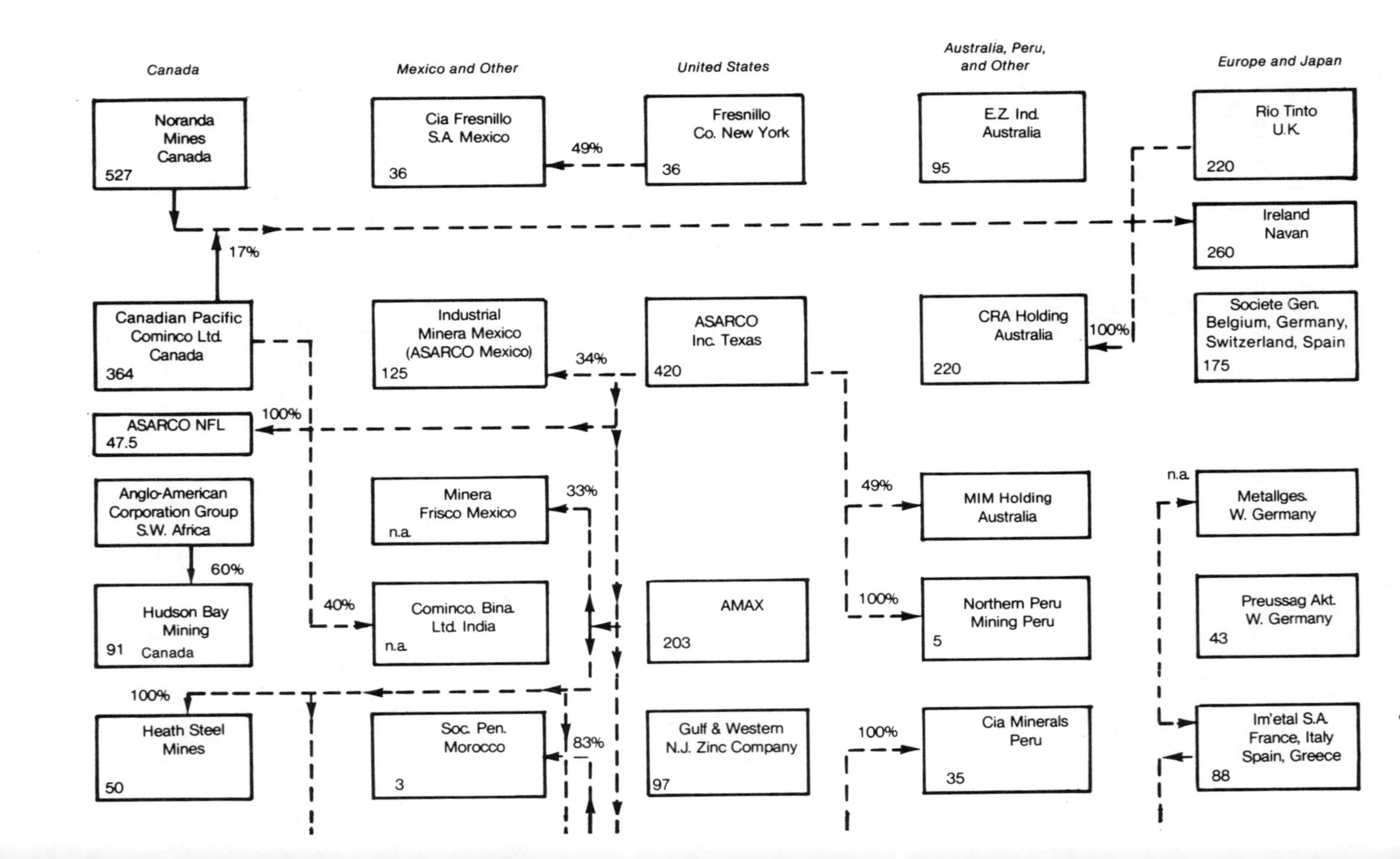
Canada
Mexico and Other
United States
Australia, Peru, and Other
Europe and Japan
Noranda Mines Canada 527
Cia Fresnillo S.A Mexico 36
49%
Fresnillo Co. New York 36
E.Z. Ind. Australia 95
Rio Tinto U.K. 220
17%
Ireland Navan 260
Canadian Pacific Cominco Ltd. Canada 364
Industrial Minera Mexico (ASARCO Mexico) 125
34%
ASARCO Inc. Texas 420
CRA Holding Australia 220
100%
Societe Gen. Belgium, Germany, Switzerland, Spain 175
ASARCO NFL 47.5
100%
Anglo-American Corporation Group S.W. Africa
Minera Frisco Mexico n.a.
33%
49%
MIM Holding Australia
n.a.
Metallges. W. Germany
60%
Hudson Bay Mining 91 Canada
40%
Cominco. Bina. Ltd. India n.a.
AMAX 203
100%
Northern Peru Mining Peru 5
Preussag Akt. W. Germany 43
100%
Heath Steel Mines 50
Soc. Pen. Morocco 3
83%
Gulf & Western N.J. Zinc Company 97
100%
Cia Minerals Peru 35
Im'etal S.A France, Italy Spain, Greece 88

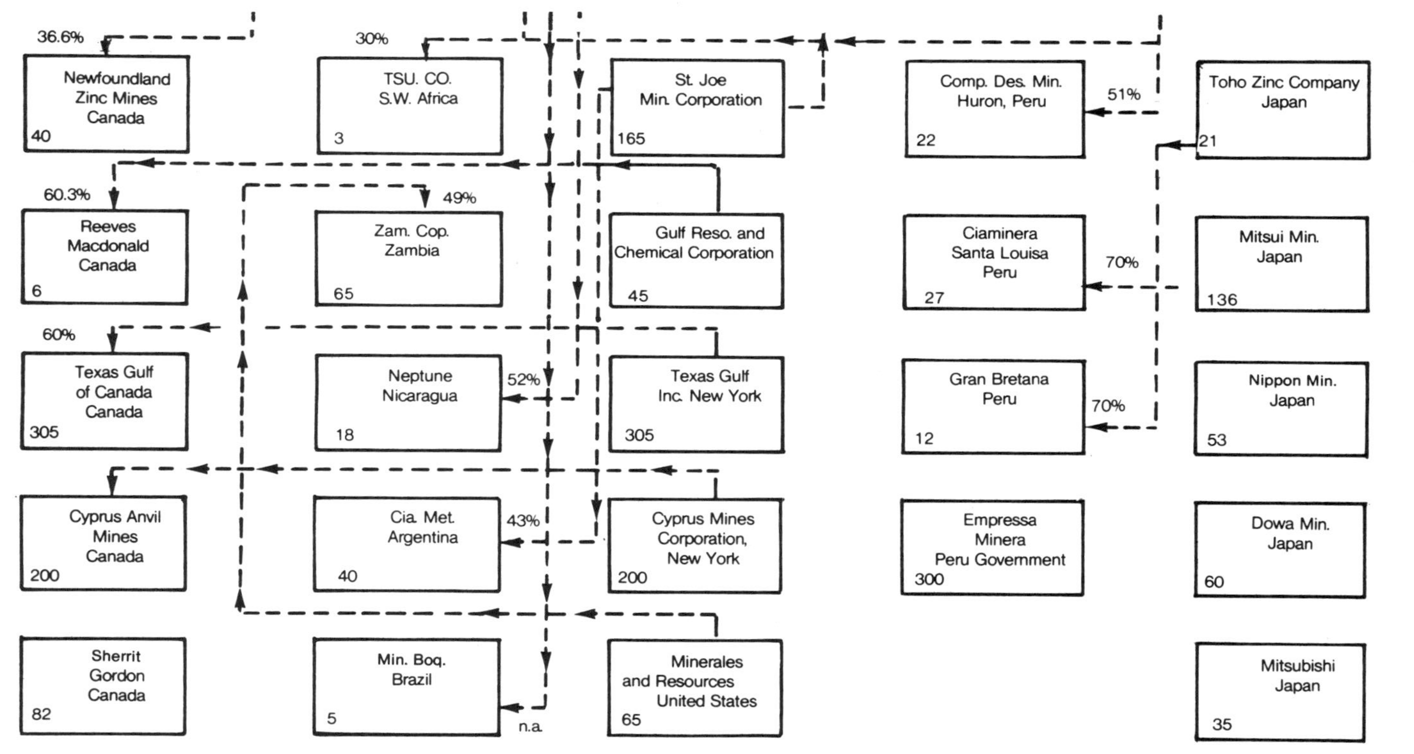

Source: Tables 3–3 to 3–7

Note: Figures inside boxes show mine production control in thousands of metric tons. Figures on the arrows indicate financial control. Coverage 80 percent of total FME world zinc mine production.

Figure 3–1. Corporate Structure of FME World Zinc Ore Production

producer and the sole metal producer in Peru. In 1973 the government of Peru, by a decree of law, expropriated the corporate assets of this company and vested the full power of the company in a newly created organization named Empressa Minera del Centro del Peru. Thus, the Peruvian industry stands as a contrast to the Australian and Mexican industries. The Peruvian output may, in fact, be treated as a one company output which represents about 9 percent of the total FME world zinc mine capacity.

Europe and Japan

European companies control about 15 percent of the total FME world zinc mine capacity, 65 percent of which lies within Europe. The only substantial control outside Europe is that of the Rio Tinto Zinc Corporation (United Kingdom) in Australia. The Rio Tinto Zinc Corporation has a controlling interest in about 45 percent of the Australian zinc mining industry. However, the mine capacity within Europe is thinly spread over many countries such as France, West Germany, Italy, Norway, Sweden, Spain, and Yugoslavia. About 40 percent of the European zinc mine capacity is controlled by four major European smelting companies (constituting three fourths of the European smelting capacity). The picture is different if one looks at the countries individually: one or two companies in each country control nearly the whole of the mining capacity in that country. One company in each of the following: Italy, Ireland, Yugoslavia, Norway, Sweden, France, and Spain controls between 80 and 100 percent of the total zinc mine capacity in that country.

The Japanese zinc mine capacity constitutes only 5 percent of the world capacity. Also, the controlling interests in terms of the number of companies are diverse. There are five Japanese companies with interests in Japanese mines. Outside interests of Japanese companies are limited to only two small mines in Peru. Although five is a large number in comparison to the number of companies in the European countries, these five companies, and many other Japanese companies, are well integrated through the banking sector.

Thus, looking at the corporate structure, the world zinc mine industry emerges as fairly well concentrated in terms of the U.S. and Canadian controlling groups. The companies based in the United States and Canada share in the control of more than two-thirds of the free-market world zinc mine production. Observing the U.S. control in the outside world, one finds it difficult to justify past U.S. protectionist policies, particularly tariffs, quotas, and stockpiles, as far as zinc mine production is concerned, since small mines were already protected through cash incentive programs. Further, it was not in the interest of the U.S. smelters to have these protectionist policies instituted as far as imports of zinc ore for their smelters were concerned.

Vertical Integration and the Corporate Structure of the Smelting Industry

The world zinc smelting industry is more concentrated than the world zinc mining industry. About 85 percent of all zinc mine production and 95 percent of all smelting capacity is controlled by the producer groups integrated to various degrees as shown in figure 3–2 and table 3–1. A large number of the big smelting companies are also integrated forward to the metal fabricating stage and backward to hydroelectric power, transport, marketing, distribution, and a few other such required facilities. Many zinc companies produce and market other products besides zinc, and also by products such as lead, silver, copper, cadmium, sulphuric acid, and fertilizer.

United States

In 1974 the United States had 600,000 tons of zinc reduction capacity. All of this capacity was controlled by six corporate groups (see table 3–1). These companies also controlled about 50 percent of the U.S. mine production. Asarco, St. Joe Minerals Corporation, and Mineral Resources Corporation also had a controlling interest of about 200,000 tons of slab zinc production capacity outside the United States. Texas-Gulf, Inc., had all its zinc mine and metal production capacities in Canada. Thus, the American companies controlled about 900,000 tons of slab zinc production capacity.

This control in 1974 had little effect, however, due to the closure of many zinc reduction plants in the United States from 1969 to 1973, as noted above. During this period, more than 600,000 tons of zinc metal production capacity was scrapped mainly due to obsolescence and higher costs of operations required by the environmental laws in the country. Closure of some reduction plants also occurred in other parts of the world but these were replaced later. The lost U.S. zinc plant capacity was not replaced due to a combination of factors such as President Nixon's Economic Stabilization Program from August 1971 to December 1973, the Environmental Protection Act of 1969, higher energy costs, and an uncertainty regarding the disposition of 1.4 million tons of zinc in the government stockpile.

If the 600,000 tons capacity that was scrapped after 1969 were included, U.S. control on zinc reduction capacity amounted to 1.4 million tons (one-third of the total FME zinc capacity), which matches very closely the U.S. control of FME world zinc mine production. Thus, in 1969 one would roughly put the United States as controller of a one-third share of mine production, smelter production, and consumption in the free-market world.

As the U.S. control on mine production and smelter capacity was almost equal, the integration in terms of national control of these two phases of the

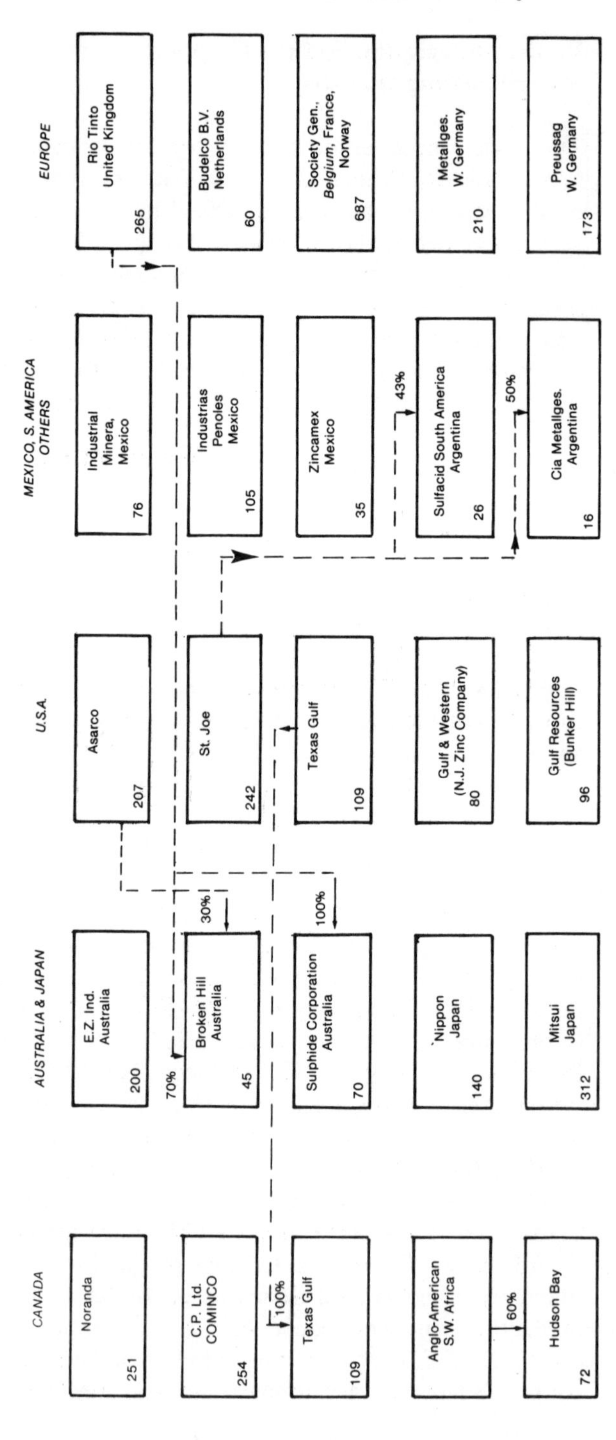

CANADA
Noranda 251
C.P. Ltd. COMINCO 254
Texas Gulf 109
Anglo-American S.W. Africa
Hudson Bay 72
100%
60%
AUSTRALIA & JAPAN
E.Z. Ind. Australia 200
Broken Hill Australia 45
Sulphide Corporation Australia 70
Nippon Japan 140
Mitsui Japan 312
70%
30%
100%
U.S.A.
Asarco 207
St. Joe 242
Texas Gulf 109
Gulf & Western (N.J. Zinc Company) 80
Gulf Resources (Bunker Hill) 96
MEXICO, S. AMERICA OTHERS
Industrial Minera, Mexico 76
Industrias Penoles Mexico 105
Zincamex Mexico 35
Sulfacid South America Argentina 26
Cia Metallges. Argentina 16
43%
50%
EUROPE
Rio Tinto United Kingdom 265
Budelco B.V. Netherlands 60
Society Gen., Belgium, France, Norway 687
Metallges. W. Germany 210
Preussag W. Germany 173

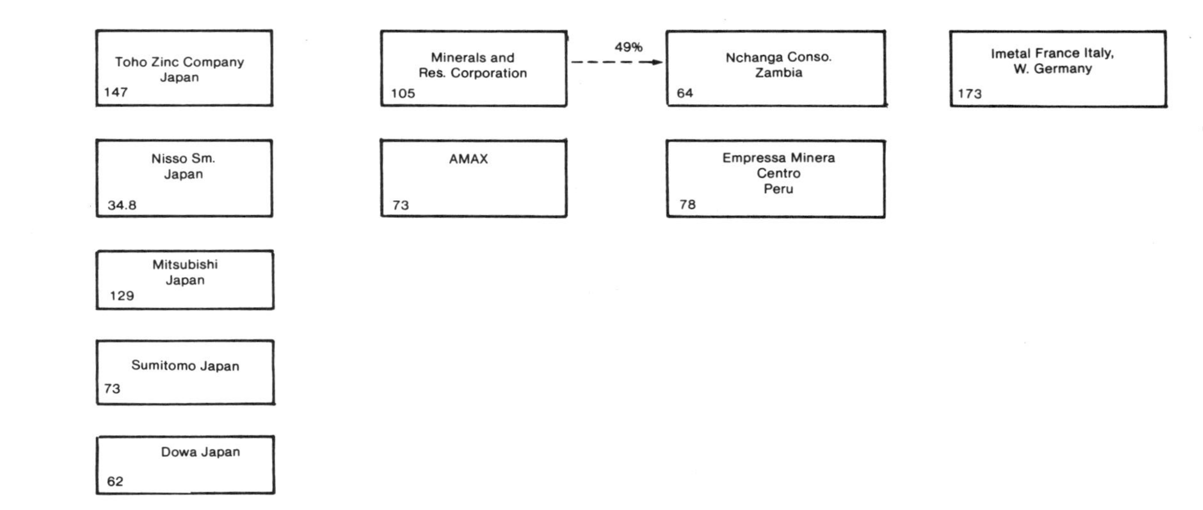

Source: Tables 3–3 to 3–7.

Note: Figures inside boxes are smelter product in thousand metric tons. Figures on arrows indicate amount of financial control.

Figure 3–2. Corporate Structure of FME World Smelter Production

industry was more or less complete. However, the control of each corporate group on mine production and metal production capacity was remarkably different. Asarco, Amax and Texas-Gulf, which together controlled 930,000 metric tons of zinc mine production capacity, controlled only about 300,000 metric tons of smelting capacity. This, however, excludes the closure of smelters before 1974. A large part of the mine production under their control in the foreign countries was smelted at the local smelters.

Canada

In Canada, smelting capacity was far below mine production capacity. Only 43 percent of the large mine production of 1.6 million tons was smelted locally. All smelter capacity in Canada (680,000 metric tons) was controlled by four corporate groups, two of which were U.S. companies. Cominco and Noranda each had a capacity close to 250,000 metric tons. Texas-Gulf and Hudson Bay controlled the rest of the smelting capacity.

Smelters in Canada, individually, controlled more than sufficient mine capacity within Canada itself. Thus, backward integration of smelters with mines in Canada was complete. Further, Cominco's backward integration extended to the ownership of more than the required hydroelectric power, shipping and dock facilities, and exploration activities. Cominco and Hudson Bay are also integrated forward for the manufacture of zinc die casts. Besides, Cominco is one of the largest producers of fertilizers for which the company profitably uses the zinc by product, sulphuric acid. Cominco and Noranda also have a wide network of trading arrangements in both Europe and North America.

Japan

The Japanese zinc smelting industry at present is leading the world, representing about 20 percent of the FME world zinc smelting capacity. In 1974 six companies in Japan controlled about 900,000 metric tons of smelting capacity. But unlike the United States and Canada, the control on mining was only one-third of its smelting capacity, Japan thus depending for the remaining two-thirds smelting capacity on foreign imports of zinc ore.

All the Japanese mine and metal production capacity is owned by domestic producers. These producers are, in turn, closely associated with the trading corporations that are responsible for purchasing and marketing all the mineral products both at home and abroad. The concentration in terms of market power is even greater due to the close association of the zinc industry through the banking sector and government agencies. An example of close cooperation between the various companies in the field of zinc smelting itself will make this more clear. The six Japanese smelting companies have pooled their resources for joint ownership of two smelting plants: Akita Smelting Company (with a present capacity of 80,000 tons, to be doubled in the near future), and Hachinohe Smelting Company (with a present capacity of 60,000 tons, to be increased to 85,000 tons). Both these smelters are operated on a toll basis for the owners on an agreed basis of consignment. This kind of arrangement allows each owner to expand his own smelters on a more economic basis.

Europe

The zinc metal industry in Europe is so highly integrated, both for production and marketing, that it becomes meaningless to analyze the concentration on a country basis (except for West Germany). Five corporate groups (see table 3–1) control about 1.5 million metric tons (about one-third of the total free-world zinc plant capacity) of metal production capacity. The Rio Tinto Zinc Corporation, which controls 265,000 metric tons of plant capacity, is the only European company with any substantial controlling interest outside Europe (115,000 metric tons in Australia). Metallgeselleschaft and Preussag are two German companies with 210,000 and 175,000 metric tons capacity. The other two corporate groups, Société Générale de Belgique and Imétal, have controlling interests with capacities of 690,000 and 175,000 metric tons, respectively, in France, Italy, Norway, Sweden, and West Germany. Preussag and Imétal are, in turn, interlinked through a common subsidiary. The concentration in Europe is further strengthened through EEC's customs union policies. These five corporate groups together control about 80 percent of the total European zinc plant capacity.[6] All these companies also have controlling interests in Europe. But, since Europe has only 600,000 tons mine capacity, the remaining 900,000 tons of foreign ore (half of the total ore traded in the world market excluding interEuropean trade) is imported. Further, some of these companies are also integrated both backward and forward like Cominco in Canada.

Australia, Mexico, and Peru

About 60 percent of the zinc produced in Australia is smelted there. There are three zinc smelters, the Electrolytic Zinc Company of Australasia (capacity 200,000 tons), the Sulphide Corporation Private Ltd. (capacity 70,000 tons), and Broken Hill Associated Smelters (capacity 45,000 tons). While the former is domestically controlled, the latter two are under the

Table 3–1
Vertical Integration
(thousand metric tons zinc content)

Country/Corporate Family (A)	*Mine Capacity (B)*	*Smelter Capacity (C)*	*Mine Deficit (Surplus) (D)*	*Degree of Integration (E) = [(C) ÷ (B) ≥ 1] = 1*	*Multinational Operations (F)*
Australia					
1. E.Z. Industries	95	200	105	1.0	
Canada					
2. Noranda	527	251			
3. Texasgulf	305	109			India (n.a.)
4. Cominco	364	254			Greenland (n.a.) Spain (n.a.) United States (n.a.)
5. Cyprus Anvil	185	—			
6. Sherrit Gordon	252	—			
7. Hudson Bay	91	72			
United States					
8. St. Joe	165	242	77	1.0	Argentina (40 mine, 42 sm.) Peru (35 mine)
9. Asarco	420	207	(213)		Australia (110 mine), Mexico (125 mine 62 sm.) Peru (5 mine), Nicaragua (18 mine),
10. Amax	203	73	(130)		Canada (75 mine), Mexico (50 mine), S. W. Africa (3 mine)
11 Gulf & Western	97	80	(17)		
12. Gulf Resources	45	96	51	1.0	Canada (6 mine)
13. Minerals	65	105	40	1.0	Zambia (65 mine, 64 sm.)
Japan					
14. Mitsui	136	312	176	1.0	Peru (27 mine)
15. Toho	21	147	126	1.0	Peru (12 mine)
16. Nippon	53	140	87	1.0	
17. Mitsubishi	35	129	94	1.0	
18. Sumitomo	—	72	72	1.0	
19. Dowa	60	62	2	1.0	

Europe					
20. Société Générale de Belgique	175	687	512	1.0	
21. Metallgesellschaft	70	210	140	1.0	
22. Imetal	88	214	126	1.0	Peru (22 mine) Brazil (5 mine) Morocco (3 mine)
23. Preussag	43	173	130	1.0	
24. Rio Tinto	220	265	45	1.0	Netherlands (60 sm.) Australia (220 mine, 115 sm.)
Total	3715	4160			

Source: Tables 3–3 through 3–7.

control of the Rio Tinto Zinc Corporation of the United Kingdom. The Broken Hill and Sulphide plants obtain sufficient zinc concentrates from the mines of associate companies; Electrolytic Zinc Co. depends on the mines controlled by other companies (mainly Broken Hill) for half of its required zinc ores (the other half being provided by its own mines).[7]

In Mexico the major smelters and refiners are Zinc Industrial Penoles, Industrial Minera Mexico, S.A. (Asarco has minor interests—34 percent—in this company), and Zincamex (government-owned). The Industrial Penoles, the largest smelter in the country, has a capacity of 65,000 tons, followed by Industrial Minera Mexico with a capacity of 62,000 tons and Zincamex with a capacity of 30,000 tons. Both Industrial Minera Mexico and Industrial Penoles acquire sufficient concentrates from their own mines and from the mines of their associate companies. Zincamex does not have its own mines but smelts concentrates from Minera Frisco and Industrial Minera Mexico mines. This total smelter capacity may be compared with the production of zinc concentrates in 1974 at 262,000 tons.

Peru has only one smelter, which is state-owned with a capacity of 78,000 metric tons. The concentrates for this smelter are provided by the former Cerro de Pasco mines and some other smaller independent mines, which leaves more than 80 percent of the zinc concentrates in the country to be exported to other smelters in the world for processing.

Two smelter plants in Argentina with a combined capacity of 32,000 tons are under the control of St. Joe Minerals Corporation of the United States. The Nehanga smelter in Zambia is partially controlled (49 percent) by the Minerals and Resources Corporation of the United States. This Zambian plant has a capacity of 64,000 tons. In both countries concentrates are provided by the local mines.

Future Developments and Corporate Control

Some foreseeable developments in corporate structure and vertical integration might provide a good guide for the analysis of the possible behavior of the world zinc market in the near future.[8] A trend has already been observed in many developing countries toward a nationalistic attitude resulting in local ownership of resources, under the control of both the state and the individual nationals of the countries concerned. A few major corporate groups have also shown a tendency toward joint ventures. Further, more vertical integration is likely to be encouraged as an announced policy of the United Nations. Clearly the implications of these developments for future market behavior could be enormous (see table 3–2).

Peru, one of the largest producers of zinc ore, has nationalized most of its mines recently. Now all minerals, besides being under state production control, are supposed to be marketed through Minero Peru—a government organization. Centromin, another government company, controls all smelter capacity in Peru. Minero Peru has recently completed, with the help of Metallgeselleschaft A.G., a feasibility study for an electrolytic zinc plant with a capacity of 89,000 tons. Once this plant is fully set up, Peru will be able to smelt about one-half of its mine production within the country.

In Mexico during 1974, Mexican interests increased their participation in Asarco Mexicana, the major zinc mining company of Mexico, by a 51 percent interest in Cia Fresnillo S.A., a New York-based company. Minera Frisco, previously owned by San Francisco Mines of Mexico (a California company) was expropriated by the Mexican government. These four companies, now holding the major Mexican interests, control practically all the zinc mining activities in Mexico. Industrial Minera Mexicana (IMM), Industrial Penoles (IP), and Zincamex are the major smelting concerns. Both IMM and IP are expected to expand their total smelting capacity to about 170,000 tons by 1976 and 215,000 tons by 1980.[9] Thus, by 1980, assuming the production of zinc concentrates will remain at present levels, Mexico will be smelting nearly all its mine production of zinc. Further, as is obvious both in Mexico and Peru, corporate control by foreign companies is giving way to public control of both mine and metal production.

Matilda, the major zinc mine in Bolivia, previously under the control of the U.S. Steel Corporation and Philipp Bros., has recently been nationalized by the Bolivian government. The New Jersey Zinc Co. of the United States and the Dowa Mining Co. of Japan are cooperating for the development of the Huari Huari zinc mine.[10] This mine is believed to have zinc ore reserves of 1.6 million metric tons with 20-percent zinc. On the other hand the Soviet and Polish interests are conducting a feasibility study for a zinc smelter with about 50,000 tons capacity.

In India, all mine production of zinc is in the public sector. The Cominco-

Table 3–2
FME World Smelter Capacity, 1974 and 1980
(thousand metric tons)

Country	*Estimated Capacity 1974*	*Percent of the FME World Capacity*	*Expected Capacity 1980*	*Percent of the FME World Capacity*	*Percent Increase 1975–1980*
America					
United States	635	12.9	859	13.5	35.3
Canada	557	11.3	786	12.3	41.1
Mexico	157	3.2	245	3.8	56.1
Argentina	51	1.0	51	0.8	0.0
Brazil	35	0.7	125	2.0	257.0
Peru	73	1.5	235	3.7	318.0
Europe					
Belgium	321	6.5	333	5.2	3.7
France	298	6.0	298	4.7	0.0
West Germany	424	8.6	459	7.2	8.2
Italy	235	4.8	300	4.7	27.6
Netherlands	140	2.8	150	2.3	7.1
U.K.	90	1.8	90	1.4	0.0
Other Europe	438	8.9	743	11.6	70.0
Africa	199	4.0	259	4.1	30.1
Australia	315	6.4	315	4.9	0.0
Asia					
Japan	898	18.3	964	15.0	7.3
India	38	.77	77	1.2	103.0
Rep. of Korea	15	.3	76	1.2	407.0
World (FME)	4,919	100.0	6,386	100.0	29.8

Source: "Zinc," *Mineral Bulletin*, MR 159 (Ottawa: Department of Energy, Mines and Resources, 1976), pp. 51–55. Reprinted with permission.

Binani Smelter, in which Cominco has a 40-percent interest, is having that interest reduced to 33 percent. The plant is being expanded from 20,000 to 40,000 tons capacity. The Bon-Becker deposits in Morocco, which were owned by Cia Asturiene des Mines S.A., were "morocconized" in 1974.

In Ireland, too, there has been a conflict between the government and private interests over the Navan Mine, the largest ever found in Europe.[11] Since the right to the deposits is owned by the state, the opinion has been expressed that the state should have a substantial share in profits from the mine, rather than simply the royalties. The Irish government wants Tara Explorations Ltd., in order to be able to obtain a mining lease, to agree to not more than 49 percent participation in the mine, to taxation at 50 percent, and royalties at 10 percent. Tara, originally a subsidiary of Northgate Explorations Ltd., now together with Northgate owns less than 50 percent of the mine. Cominco and Charter Consolidated own 31 percent, and Noranda Mine another 20 percent.[12] Part of the concentrates from the mine is expected to be smelted at the Hartlepool plant (owned by Cominco) and the rest at a newly proposed electrolytic zinc plant (60,000 tons capacity) jointly owned by Tara and Noranda.

Argentina, Greenland, and Iran are relatively new entrants into the world zinc industry. Government influence has not been extended in these countries because the technological expertise and/or financial resources are lacking. Most of Argentina's zinc concentrate and slab zinc are produced by a subsidiary of the St. Joseph Lead Co. of the United States. The Black Angel Mine in Greenland, with an estimated reserve of 5 million tons (20-percent zinc) and a current rate of production of about 60,000 tons, is controlled by Cominco (major interest) and Northgate Explorations Ltd. Simiran, an Iranian company, with the help of Rio Tinto (34 percent) and Pennaroya (17 percent) had proposed to prospect zinc mine production in Iran. Since the shares of these two foreign companies equal 51 percent, they are likely to exercise control over production. Meanwhile, the government of Iran is contemplating setting up a smelter plant for smelting domestically produced zinc ore locally in the near future.

Zaire and Zambia are the only two major African producers of zinc. Zaire, the largest mine producer of zinc (156,000 tons in 1973) smelts less than half of its zinc concentrates. Zambia produces about 65,000 tons of zinc ore annually, and all these concentrates are locally smelted. The production of zinc in this country is controlled by the Anglo-American Corporation and the Minerals and Resources Corporation of the United States. Recently it has been proposed that the smelting capacity in Zambia be enlarged to about 115,000 tons, which would imply that part of the exports of zinc ore from Zaire to Europe may be diverted to Zambian smelters.

The changes in the structure of ownership and integration in Australia and Canada, the largest producers of raw zinc, are of substantial interest to the world zinc industry. In Australia, as noted above, three-quarters of the mine production of zinc and about one-third of the smelting capacity is controlled by European and American interests. Recently many new zinc deposits have been discovered in Australia by American controlling interests. Jododex Australia Pty. Ltd., jointly owned by the St. Joe Minerals Corporation and the Phelps Dodge Corporation, have discovered a new zinc deposit at Woodlawn with estimated reserves of 7 million metric tons (9.4 percent zinc, 3.3 percent lead, 2.9 percent copper, and 1.9 oz/ton silver) which was scheduled to start production some time in the mid 1970s. Similarly, Mount Isa Mines Ltd., a wholly owned subsidiary of M.I.M. Holdings, has prospected 35.6 million metric tons of reserves at the Hilton Mine (9.6 percent zinc, 7.7 percent lead, 6.3 oz/ton silver) and a 200 million metric ton lead-zinc deposit in the MacArthur River area.[13] In western

Australia, the Electrolytic Zinc Company has joined Amax in prospecting the Golden Grove copper-zinc deposit with a very high zinc value (24 percent). Similarly, Conwest Explorations Ltd., a Canadian company, has reported a possibility of a large zinc deposit with a very high zinc value (about 23 percent). Reserves of these deposits are yet unknown. Placer Prospecting Pty. Ltd. (Australian), where controlling interests are Placer Development Ltd. (Australian), Kaiser Aluminum and Chemical Corporation (American), and Traiko Pty. Ltd. (Australian) have prospected the Lady Loretta deposit in Queensland. The deposit is estimated to have 9 million metric tons of ore, grading 18 percent zinc, 7 percent lead and 3.5 oz/ton of silver. Thus, Australia seems a very promising source of future supply for zinc, though controlling interests are not only local, but also belong to the well-established major zinc producers elsewhere.

Not much is known regarding the plans for expansion of the smelting capacity in Australia. However, given the stricter environmental regulations in the United States, it is more likely that the American-based companies associated with the production of zinc ore would prefer to smelt the ores within Australia, unless they decide to toll smelt in some other countries.

The Canadian zinc industry in the past has tended to become more concentrated in terms of the corporate structure. Cominco and Noranda have acquired assets of many smaller companies in the past. Further, these companies have come together in some ventures such as Tara Explorations Ltd., indicating the possibility of cooperation in terms of acquiring market power. More than 80 percent of the resources to be utilized for production of zinc in the future lie under the control of the "big six" companies. New developments, given the tendencies of the past, are more likely to perpetuate the present corporate structure.

The zinc smelting industry in Canada is believed to have plans for substantial development in the near future. During the five-year period 1975–1980, the smelting capacity in Canada is expected to increase from about 550,000 metric tons to about 790,000 metric tons (an increase of about 68 percent). By the year 2000, the present estimate envisages an increase in smelting capacity to about 1.5 million metric tons. Comparing these estimated mine production figures of about 1.25 million metric tons in 1974, 1.4 million metric tons in 1980, and about 2.5 million by the year 2000, there appears to be a conscious effort to integrate mine production with smelting capacity within the country at least in the first few years.[14] Still, Australia and Canada will probably remain the major suppliers of zinc ore in the world market.

Government interference and nationalistic tendencies in Canada, in general, although in an embryonic stage at present, are developing quite fast. In 1973 the Canadian Development Corporation acquired a minority controlling interest in Texas-Gulf Inc. CDC and other Canadian shareholders now own 42 percent of the company's assets. Four of the twelve representatives on the Board of Directors are from CDC. The provincial governments of British Columbia, Manitoba, and Ontario have attempted to raise taxes, one of the important current issues in Canadian mining.

Both in Japan and the United States the current corporate structure is likely to continue. Corporate groups in both countries are, however, trying to expand their control in other countries, particularly for raw zinc, as we noted previously. During the period 1975–1980, no expansion of smelting capacity in Japan is anticipated. In the United States, the smelting capacity is expected to increase from 625,000 metric tons in 1975 to 860,000 metric tons in 1980. This increase in capacity in the United States is, however, only a partial replacement of the smelter capacities scrapped during the 1969–1973 period. An interesting feature of the increase in capacity during the 1975–1980 period is the joint venture of Asarco and M.I.M. Holdings of Australia and that of the New Jersey Zinc Company with the Union Minière S.A. It seems that the main objective of these American corporations is to secure steady sources of raw zinc for their smelters. The New Jersey Zinc Company is also planning to buy antipollution technology from the Dowa Mining Co. of Japan.

International Organizations

Besides the structure of ownership, the existence of some national and international organizations can play a vital role in coordination of the market. OPEC, CIPEC, and the International Tin Association are glaring examples in the mineral commodity markets.

Although there has been no formal cartel in the world zinc industry since 1935, producers of zinc have come together for the realization of various goals.[15] The International Lead and Zinc Study Group (ILZSG), the International Lead and Zinc Research Organization (ILZRO), and the American Zinc Institute (AZI) are the major organizations in the zinc industry. Besides, the major zinc producers have also come together for determining market policies through various meetings and conferences initiated by the United Nations.

International Lead and Zinc Study Group

The conditions leading to formation of the ILZSG reveal the goal of this organization. Following the Korean War boom, the instability of commodity prices and the consequent erosion of the foreign-exchange earnings of the less-developed countries drew the attention of the United Nations. The

United Nations provided a forum for the producers of primary commodities in which to discuss their problems. An interest in the program was also shown by the major consumers, to avoid instability in the commodity markets. An Interim Coordinating Committee for International Commodity Agreements (ICCICA) was established under the auspices of the Economic and Social Council (ESC) of the United Nations. ICCICA initiated the formation of study groups for primary commodities. However, the program was not pushed ahead until 1957, when many zinc mine closures alarmed the major zinc-producing countries. The formation of ILZSG was approved in late 1958 and the study group was formally established in 1960. At present the number of members has increased to more than 30 countries.[16] The Group holds regular meetings in the fall of every year.

The major aims as outlined in the constitution of the Group are:

> to provide the opportunities for appropriate intergovernmental consultations on international trade in lead and/or zinc and make such studies of the world situation in lead and zinc as it sees fit, having regard especially to the desirability of providing accurate information regarding the supply and demand position and of its probable development.... The Group may report to Member Governments, such reports may include suggestions and/or recommendations.[17]

ILZSG, through its monthly publication and some occasional research publications, has made substantial improvements in the availability of individual country statistics on mine production, primary and secondary metal production, consumption, and relatively aggregative information on international trade, stocks, and prices. Further, the Statistical Committee of the Group has helped the member nations through forecasting probable developments in demand and supply in various major consumer and producer countries of the world. Finally, a rapport has been encouraged and developed between major zinc-producing companies and countries leading to an exchange of ideas and more realistic forecasts of supply-demand balance and the general outlook of the industry as a whole. The relative price stability in the world zinc market for as long as two and one-half years at a time during the 1958–1972 period may be, to some extent, credited to the efforts of the Group.

The International Lead Zinc Research Organization Inc. (ILZRO)

ILZRO, established in September 1958 as the Expanded Research Program of the Lead Industries Association and American Zinc Institute, gained its international character in 1963 to reflect the worldwide sponsorship of the

Table 3–3
Structure of the Canadian Zinc Industry

Major Corporate Family	*Family Control (% shares)*	*Primary Zinc Production Facilities*	*1974 Capacity (tons of zinc)*	
			Mine	*Refinery*
Noranda Mines Ltd.				
Geco Mines Ltd., Manitouwadge	100.0	copper-zinc-silver mine	78,000	—
Brunswick Mining & Smelting Corp. Ltd. Bathhurst, N.B.	64.2	zinc-lead-silver mine	222,000	47,000
Canadian Electrolytic Zinc Ltd. Valleyfield, Que.	associate	electrolytic refinery	—	204,000
Kerr Addison Mines Ltd. Normetal, Que.	—do—	zinc-copper-silver mine	12,700	—
Mattagami Lake Mines Ltd. Mattagami, Que.	—do—	—do—	95,800	—
Mattabi Mines Ltd. Sturgen Lake, Ontario	—do—	—do—	87,700	—
Orchan Mines Ltd.	—do—	—do—	30,000	—
		Total Capacity	526,600	251,200
		Mine surplus or (deficit)	275,400	
Canadian Pacific Railways Ltd.		transport company		
Canadian Pacific Investments Ltd.	100.0	investment company		
Cominco Ltd., Trail, B.C.	100.0	electrolytic refinery	—	254,000
Sullivan Mine, B.C.	100.0	zinc-lead-silver mine	149,000	—
H.B. Mine, B.C.	100.0	—do—	14,700	—
Pine Point Mine Ltd. N.W.T.	69.0	—do—	192,700	—
Cominco-Binani Ltd., India	40.0	—do—	n.a.[a]	n.a.
Mitsubishi-Cominco Smelting Ltd. Japan	45.0	electrolytic refinery	n.a.	n.a.
		Total capacity	356,000	254,000

		Mine surplus or (deficit)	102.000	
Texas Gulf of Canada Ltd. Ontario	See United States	zinc-lead-silver mine		
		electrolytic refinery	305,300	108,900
		Mine surplus or (deficit)	196,400	
Cyprus Anvil Mining Corp.	See United States	zinc-lead-silver mine	185,400	—
		Mine surplus or (deficit)	185,400	
Sherrit Gordon Mines Ltd. Ontario				
Fox Mine, Lynn Lake	100.0	zinc-copper-silver mine	19,600	—
Ruttan Mine, Ruttan Lake	100.0	—do—	55,700	—
		Total capacity	75,300	—
		Mine surplus or (deficit)	75,300	
Anglo-American Corporation S.W. Africa				
Hudson Bay Mining and Smelting Co. Ltd.				
Flin Flone	100.0	electrolytic refinery	—	71,700
Anderson Lake				
Chisel Lake				
Osborne Lake				
Dickiston Lake	100.0	zinc-copper-silver mine	90,900	—
Schist Lake				
Stall Lake				
Ghost Lake				
		Total capacity	90,900	71,700
		Mine surplus or (deficit)	19,200	

Source: "Zinc," *Mineral Bulletin* MR 159 (Ottawa: Department of Energy, Mines and Resources 1976), pp. 1–10 and *Moody's Industrial Manual* (New York: Moody's Investors Service Inc., 1976).

[a]n.a. = not available.

Table 3–4
Structure of the Australian Zinc Industry

Major Corporate Family	*Family Control (% shares)*	*Primary Zinc Production Facilities*	*1974 Capacity (tons of zinc)*	
			Mine	*Refinery*
E.Z. Industries Ltd.				
Electrolytic Zinc Company of Australasia Ltd.	100.0			
Risdon, Tasmania		electrolytic refinery	—	200,000
Rosbery, Tasmania		zinc-lead Westcoast mines	75,000	—
Baltana, South Australia		zinc mines	20,000	—
		Total capacity	95,000	200,000
		Mine surplus or (deficit)	(105,000)	
M.I.M. Holdings, Ltd.				
Mount Isa Mines Ltd.	100.0			
Mount Isa, N. Queensland		zinc-lead mine	110,000	—
North Broken Hill Ltd.				
Broken Hill, N.S.W.		zinc-lead mine	50,000	—
Broken Hill Associated Smelters Pty. Ltd.	30.0	controlled by Rio Tinto Zinc Corp.		
Port Pirie		electrolytic refinery	—	45,000
The Rio Tinto Zinc Corp.				
CRA Holdings Pty. Ltd.	100.0			
Conzinc of Rio Tinto of Australia Ltd.	80.6			
Australiaan Mining and Smelting Ltd.	73.5			
		Total capacity	220,000	265,000
		Mine surplus or (deficit)	(45,000)	

Source: "Zinc," *Mineral Bulletin*, MR 159 (Ottawa: Department of Energy, Mines and Resources, 1976, p. 30. Reprinted with permission.

Table 3–5
Structure of the Japanese Zinc Industry

Major Corporate Family	*Family Control (% shares)*	*Primary Zinc Production Facilities*	*1974 Capacity (tons of zinc)*	
			Mine	*Refinery*
Mitsui Mining and Smelting Co. Ltd.				
Miike, Japan		vertical retort smelter	—	118,000
Miike, Japan		electrolytic refinery	—	20,000
Kamioka, Japan		electrolytic refinery	—	61,000
Hiroshima, Japan		electrolytic refinery	—	66,000
Shikama, Japan		zinc-lead Kamioka mine	80,000	
Cia Minera Santa Louisa S.A.	n.a.	zinc-lead-copper mine	27,000	—
Huanzala, Peru				
Iwami Mining Co. Ltd.	100.0			
Shimane, Japan		zinc-lead mine	3,000	—
Akita Smelting Co.[a]	10.0			
Iijima, Japan		electrolytic refinery	—	9,000
Hachinohe Smelting Co.[a]	50.0			
Hachinohe, Japan		imperial smelter	—	38,000
Nippon Zinc Mining Co. Ltd.	99.0			
Fukui, Japan		zinc-lead Nakatatsu mine	26,000	—
		Total capacity	136,000	312,000
		Mine surplus or (deficit)	(176,000)	
Toho Zinc Co.				
Annaka, Japan		electrolytic refinery	—	139,000
Taishu, Japan		zinc-lead mine	9,000	—
Gran Bretana. SMRL				
Gran Bretana SMRL	70.0			
Gran Bretana. Peru		zinc mine	12,000	—

Table 3–5 *(continued)*

Major Corporate Family	*Family Control (% shares)*	*Primary Zinc Production Facilities*	*1974 Capacity (tons of zinc)*	
			Mine	*Refinery*
Akita Smelting Co.	5.0			
Iijima, Japan		electrolytic refinery	—	4,500
Hachinohe Smelting Co.	5.0			
Hachinohe, Japan		imperial smelter	—	3,800
		Total capacity	21,000	147,300
		mine surplus or (deficit)	(126,300)	
Dowa Mining Co. Ltd.				
Akita, Japan		zinc-lead Uchinotai mine	24,000	—
Akita, Japan		zinc-lead Matsumine mine	24,000	—
Akita, Japan		zinc-lead Koyashiki mine	7,000	—
Akita, Japan		zinc-lead Fukazawa mine	5,000	—
Akita Smelting Co.	52.0			
Iigima, Japan		electrolytic refinery	—	46,800
Hachinohe Smelting Co.	20.0			
Hachinohe, Japan		imperial smelter	—	15,200
		Total capacity	60,000	62,000
		Mine surplus or (deficit)	(2,000)	
Nisso Smelting Co.				
Aizu, Japan		electrolytic refinery	—	31,000
Hachinohe Smelting Co.	5.0			
Hachinohe, Japan		imperial smelter	—	3,800
		Total capacity	—	34,800
		Mine surplus or (deficit)	(34,800)	

Nippon Mining Co. Ltd.				
Mikkaichi, Japan		electrothermic smelter	—	120,000
Ibaragi, Japan		copper-zinc Hitachi mine	4,000	—
Akita, Japan		copper-zinc Shakanai mine	11,000	—
Hokkaido, Japan		zinc-lead Toyoha mine	38,000	—
Akita Smelting Co.	14.0			
Iijima, Japan		electrolytic refinery	—	12,600
Hachinohe Smelting Co.	10.0			
Hachinohe, Japan		imperial smelter	—	7,600
		Total capacity	53,000	140,200
		Mine surplus or (deficit)	(87,200)	
Mitsubishi Metal Corp.				
Akita, Japan		electrolytic refinery	—	97,000
Hosokura, Japan		electrolytic refinery	—	20,000
Hyogo, Japan		zinc-copper Akenobe mine	6,000	—
Akita, Japan		copper-zinc Furutobe mine	4,000	—
Akita, Japan		zinc-lead Hosokura mine	20,000	—
Akita, Japan		copper-zinc Matsuki mine	2,000	—
Oppu Mining Co. Ltd.	100.0			
Aomori, Japan		zinc-lead Oppu mine	2,000	—
Yamagata, Japan		zinc-lead Yatant mine	1,000	—
Akita Smelting Co.	5.0			
Iijima, Japan		electrolytic refinery	—	4,500
Hachinohe Smelting Co.	10.0			
Hachinohe, Japan		imperial smelter	—	7,600
		Total capacity	35,000	129,100
		Mine surplus or (deficit)	(94,100)	
Sumitomo Metal Mining Co. Ltd.				
Sumiko I.S.P. Co. Ltd.	55.0			
Harima, Japan		imperial smelter	—	60,000
Akita Smelting Co.	14.0			
Iijima, Japan		electrolytic refinery	—	12,600
		Total capacity	—	72,600
		Mine surplus or (deficit)	(72,600)	

Source: "Zinc," *Mineral Bulletin*, MR 159 (Ottawa: Department of Energy, Mines and Resources, 1976), pp. 24–26. Reprinted with permission.

[a]Pro rata ownership capacity

Table 3–6
Structure of the European Zinc Industry

Major Corporate Family	*Family Control (% shares)*	*Primary Zinc Production Facilities*	*1974 Capacity (tons of zinc)*	
			Mine	*Refinery*
Société Génerale de Belgique				
Société Génerale des Minerais	77.6	commerical company		
Union Minière	36.0	investment company		
Metallurgie Hoboken Overpelt	61.5			
Overpelt, Belgium		electrolytic refinery	—	80,000
Société de Prayon	44.6			
Enhein, Belgium		electrolytic refinery	—	65,000
Société de Mines et Fonderies de Zinc de la Vieille Montagne	28.0			
Balen, Belgium		electrolytic refinery	—	168,000
Viviez, France		electrolytic refinery	—	94,000
AG des Altenbergs fur Bergbau und Zinkhuttenbetrieb	100.0			
Luderich, Germany		zinc-lead mine	15,000	—
Bolaget Vieille Montagne	100.0			
Ammesberg, Sweden		zinc-lead mine	25,000	—
Compagnie Royale Asturienne des Mines	25.9			
Auby, France		vertical retort smelter	—	90,000
Santander, Spain		zinc-lead Reocin mine	45,000	—
Boliden Aktiebolag	11.0			
Stockholm, Sweden		zinc-lead-copper mines	75,000	—
Det Norske Zinkkompani	50.0			
Odda, Norway		electrolytic refinery	—	85,000
Austurianna de Zinc	50.0			
Aviles, Spain		electrolytic refinery	—	105,000
Guipozcoa, Spain		zinc-lead mine	15,000	—
		Total capacity	175,000	687,000
		Mine surplus or (deficit)	(512,000)	

Metallgesellschaft AG				
Berzelius Metallhutten Gesellschaft GmbH	100.0			
Duisberg, Germany		imperial smelter	—	80,000
Ruhr-Zinc GmbH	100.0			
Datteln, Germany		electrolytic refinery	—	130,000
Sachtleben Aktiengesellschaft fur Bergbau GmbH	100.0			
Lennestadt, Germany		zinc-lead Meggen mine	55,000	—
Ramsbeck, Germany		zinc-lead Ramsbeck mine	15,000	—
		Total capacity	70,000	210,000
		Mine surplus or (deficit)	(140,000)	
Imétal S.A.				
Compagnie de Mokta	93.8	investment company		
Compagnie des Mines de Huaron	51.0			
Huaron, Peru		zinc-lead mine	22,000	—
Societé Minière et Métallurgique de Peñarroya	58.0			
Noyelles Godault, France		imperial smelter	—	105,000
Herault, France		zinc-lead Malines mine	10,000	—
Ardeche, France		zinc-lead Largentiere mine	4,000	—
Societa Mineraria e Metallurgica di Pertusola	75.7			
Crotone, Italy		electrolytic refinery	—	82,000
San Pietro di Cadore, Italy		zinc-lead Salalossa mine	22,000	—
Preussag-Weser-Zink GmbH[a]	25.0			
Nordenham, Germany		electrolytic refinery	—	27,500
Compagnie Française des Mines du Laurium	66.6			
Laurium, Greece		zinc-lead mine	2,000	—
Sociedad Minera y Metallurgica de Peñarroya España S.A.	98.1			
Carthagena, Spain		zinc-lead mine	20,000	—
Sociéte Peñarroya-Maroc	82.8			
Morocco		lead-zinc mine	3,000	—
Mineraçao Boquira S.A.	n.a.			
Bahia, Brazil		lead-zinc mine	5,000	—
		Total capacity	88,000	214,500
		Mine surplus or (deficit)	(126,500)	

Table 3–6 *(continued)*

Major Corporate Family	*Family Control (% shares)*	*Primary Zinc Production Facilities*	*1974 Capacity (tons of zinc)*	
			Mine	*Refinery*
Pressag Aktiengesellschaft				
Harlingerode, Germany		vertical retort smelter	—	94,000
Goslar, Germany		zinc-lead Rammelsberg mine	30,000	—
Bad Grund, Germany		zinc-lead Grund mine	13,000	—
Preussag-Weser-Zink GmbH	75.0			
Nordenham, Germany		electrolytic refinery	—	79,000
		Total capacity	43,000	173,000
		Mine surplus or (deficit)	(130,000)	
The Rio Tinto Zinc Corp. Ltd.				
CRA Holdings Pty. Ltd.	100.0	investment company		
Conzinc Rio Tinto of Australia Ltd.	80.6	investment company		
Australian Mining and Smelting Ltd.	73.5	investment company		
AM & S Europe Ltd.	100.0	commercial company		
Commonwealth Smelting Ltd.	100.0			
Avonmouth, United Kingdom		imperial smelter	—	90,000
Australian Overseas Smelting Pty. Ltd.	100.0	investment company		
Budelco B.V.[a]	50.0			
Budel, Netherlands		electrolytic refinery	—	60,000
New Broken Hill Consolidated Ltd.	100.0			
Broken Hill, N.S.W., Australia		zinc-lead mine	140,000	—
The Zinc Corp. Ltd.	100.0			
Broken Hill, N.S.W., Australia		zinc-lead mine	80,000	—
Sulphide Corp. Pty. Ltd.	100.0			
Cockle Creek, Australia		imperial smelter	—	70,000
Broken Hill Associated Smelters Pty. Ltd.	70.0			
Port Pirie, Australia		electrolytic refinery	—	45,000
		Total capacity	220,000	265,000
		Mine surplus or (deficit)	(45,000)	

Source: "Zinc," *Mineral bulletin*, MR 159 (Ottawa: Department of Energy, Mines and Resources, 1976), p. 21–23. Reprinted with permission.

[a]Pro rata ownership capacity.

Table 3–7
Structure of the U.S. Zinc Industry

Major Corporate Family	*Family Control (% shares)*	*Primary Zinc Production Facilities*	*1974 Capacity (tons of zinc)*	
			Mine	*Refinery*
St. Joe Minerals Corporation				
Monaca, Pa., United States		electrothermic smelter	—	200,000
Missouri, United States		lead-zinc Bushy Creek mine	2,000	—
Missouri, United States		lead-zinc Hether mine	2,000	—
Missouri, United States		lead-zinc Indian Creek mine	2,000	—
Missouri, United States		lead-zinc Viburnum mine	4,000	—
New York, United States		zinc-lead Balmat-Edwards mine	80,000	—
Compania Minera Aguilar South America	99.9			
Argentina		zinc-lead mine	40,000	—
Sulfacid South America	50.0			
Borghi, Argentina		electrolytic refinery	—	26,000
Cia. Metálurgica Austral-Argentina South America	43.0			
Comodoro, Argentina		electrothermic smelter	—	16,000
Cia Minerales Santander Inc.	100.0			
Santandu, Peru		zinc-lead mine	35,000	—
		Total capacity	165,000	242,000
		Mine surplus or (deficit)	(77,000)	
Minerals and Resources Corporation Ltd.				
Prairie Investments Ltd.	100.0	investment company		
Englehard Minerals and Chemicals Corp.	30.5			
Bartlesville, Oklahoma, United States		horizontal retort smelter	—	41,000

Table 3–7 *(continued)*

Major Corporate Family	*Family Control (% shares)*	*Primary Zinc Production Facilities*	*1974 Capacity (tons of zinc)*	
			Mine	*Refinery*
Zambia Copper Investments	49.98	investment company		
Nehanga Consolidated Copper Mines Ltd.	49.0			
Broken Hill, Zambia		imperial smelter	—	34,000
Kabwe, Zambia		electrolytic refinery	—	30,000
Broken Hill, Zambia		zinc-lead mine	65,000	—
		Total capacity	65,000	105,000
		Mine surplus or (deficit)	(40,000)	
Amax Inc.				
Amax Lead and Zinc Inc.	100.0			
Sauget, Illinois, United States		electrolytic refinery	—	73,000
Heath Steele Mines Ltd.	100.0	mine operator		
Little River Joint Venture	75.0			
Newcastle, N.B., Canada		zinc-lead mine	35,000	—
Amax Lead Company of Missouri	50.0			
Boss, Missouri, United States		zinc-lead Buick mine	60,000	—
Newfoundland Zinc Mines Limited	36.6			
Daniels Harbour, N.B., Canada		zinc mine (commences 1975)	40,000	—
Minera Frisco, South America	33.0			
Minera San Francisco Del Oro	100.0			
Chihuahua, Mexico		zinc-lead-copper mine	50,000	—
Tsumeb Corporation Ltd.	29.6			
South-West Africa		zinc-lead mine	3,000	—
		Total capacity	188,000	73,000
		Mine surplus or (deficit)	115,000	
Gulf & Western Industries Inc.				
The New Jersey Zinc Company	100.0			
Palmertown, Pa., United States		vertical retort smelter	—	80,000
Gilman, Colorado, United States		zinc-lead Gilman mine	22,000	—
Ogdensburg, N.J., United States		zinc Sterling mine	30,000	—

Center Vally, Pa., United States		zinc Freidensville mine	15,000	—
Jefferson City, Tenn. United States		zinc Jefferson mine	13,000	—
Austinville, Virginia, United States		zinc-lead Austinville mine	17,000	—
		Total capacity	97,000	80,000
		Mine surplus or (deficit)	17,000	
Gulf Resources and Chemical Corp.				
Bunker Hill Co.	100.0			
Kellogg, Idaho, United States		electrolytic refinery	—	96,000
Kellogg, Idaho, United States		zinc-lead mine	21,000	—
Star Morning Unit Joint Venture	70.0			
Burke, Idaho, United States		zinc-lead mine	12,000	—
Pend Oreille Mines and Metals Co.	100.0			
Metaline Falls, Washington, United States		zinc-lead mine	6,000	—
Reeves MacDonald Mines Ltd.	60.3			
Remac, B.C., Canada		zinc-lead mine (closed 1975)	6,000	—
		Total capacity	45,000	96,000
		Mine surplus or (deficit)	(51,000)	
ASARCO Incorporated				
Corpus Christie, Texas, United States		electrolytic refinery	—	95,000
Amarillo, Texas, United States		horizontal retort (closed 1975)	—	50,000
New Mexico, United States		zinc-lead Ground Hog mine	15,000	—
Tennessee, United States		zinc Immel mine	10,000	—
Tennessee, United States		zinc Young mine	5,000	—
Tennessee, United States		zinc New Market mine	20,000	—
Newfoundland, Canada		zinc-lead Buchans mine	25,000	—
Blackcloud Joint Venture	50.0			
Leadville, Colorado, United States		zinc-lead mine	15,000	—
Northern Peru Mining Corporation	100.0			
Quiruvilla, Peru		copper-zinc-lead mine	5,000	—
M.I.M. Holdings Ltd.	49.0	investment company		
Mount Isa Mines Ltd.	100.0			
Mount Isa, Australia		zine-lead-copper mine	110,000	—
United Park City Mines Co.	16.5			
United Park City, Utah, United States		zinc-lead mine	28,000	—
Neptune Mining Company	51.8			
Vesubio, Nicaragua		zinc-lead mine	18,000	—

Table 3–7 *(continued)*

Major Corporate Family	*Family Control (% shares)*	*Primary Zinc Production Facilities*	*1974 Capacity (tons of zinc)*	
			Mine	*Refinery*
Industrial Minera Mexico South America	34.0			
Rosita, Mexico		horizontal retort smelter	—	62,000
Charcas, Mexico		zinc-lead-copper mine	18,000	—
Pappai, Mexico		zinc-lead-copper mine	17,000	—
San Martin, Mexico		zinc-copper mine	18,000	—
Santa Barbara, Mexico		zinc-lead-copper mine	32,000	—
Plomosas, Mexico		zinc-lead mine	20,000	—
Santa Eulalia, Mexico		zinc-lead mine	5,000	—
Taxco, Mexico		zinc-lead mine	15,000	—
		Total capacity	376,000	207,000
		Mine surplus or (deficit)	169,000	

Source: "Zinc," *Mineral Bulletin*, MR 159 (Ottawa: Department of Energy, Mines and Resources, 1976), pp. 26–29. Reprinted with permission.

major lead and zinc producers. In 1969 it had thirty-six members in twelve countries. The basic objectives of ILZRO are to develop new knowledge of lead and zinc through fundamental research, to improve existing products and uses of lead and zinc, and to create new products and communicate the resulting knowledge. The research is usually carried out on a contract basis in selected laboratories throughout the world.

The Zinc Institute Inc.

The American Zinc Institute played a vital role in organizing the U.S. zinc producers during the interwar period of cartelization and providing platforms to emphasize demand for protectionist policies later.[18] Since 1968 the institute has changed its name to The Zinc Institute Inc. and has opened its doors for the membership of foreign producers. The primary objective of this organization is the promotion and market-development of zinc and zinc products.

Notes

1. See, for example, Scherer (1970).
2. For details, see the appendix to this chapter.
3. The information on these aspects used in this study is gathered from a number of sources. The major sources are: *Moody's Industrial Manual* (1975), Department of Energy, Mines and Resources (1974), Roskill (1974), and Cammarota (1975), unless otherwise indicated.
4. For example, see Kreinin and Finger (1976) and UNCTAD (1974).
5. The investigation into the boards of directors of these companies, however, has not revealed any interlocking. See *Moody's Industrial Manual* (1976).
6. The remaining 20 percent is state-owned. Ammi Spa in Italy, Belberger Bergwerks Union in Austria, Española del Zinc in Spain, Outukumpu Oy in Finland and the Trepa, Zletovo and Zorea plants in Yugoslavia are state organizations.
7. Recently, this refinery has undergone considerable expansion through the introduction of a new Jerosite process whereby recovery of zinc from concentrates could increase by another 10 percent.
8. For detailed discussion of some major aspects, see Roskill (1974, chapter 4).
9. IP is planning to increase its capacity from 105,000 tons in 1976 to 200,000 tons ultimately. At that scale, the plant will also produce 180,000 tons of sulphuric acid and 850 tons of cadmium, and thus an estimated total gain of U.S. $160,000 a day in foreign exchange for Mexico.

10. The latter extending a loan of $3.5 million to the former in return for a 10-year supply of zinc ore at the rate of 50,000 metric tons per year (containing 26,000 metric tons of zinc).

11. The mine is believed to have 77 million tons of zinc ore reserves containing 11 percent zinc and 2.62 percent lead.

12. It is believed that Cominco wanted control of the mine to provide concentrate for its new smelter at Hartlepool in the United Kingdom.

13. Asarco of the United States has 49 percent controlling interests in M.I.M. Holdings.

14. See Department of Energy, Mines and Resources (1976, pp. 51–55).

15. For a history of cartelization, see the appendix to this chapter.

16. On 26 September 1969, the governments of 30 countries—Algeria, Australia, Austria, Belgium, Bulgaria, Canada, Czechoslovakia, Denmark, Finland, West Germany, Hungary, India, Italy, Japan, Mexico, Morocco, Netherlands, Norway, Peru, Poland, S. Africa, Spain, Sweden, Tunisia, United Kingdom, United States, USSR, Yugoslavia, Zaire, and Zambia—were the members of ILZSG.

17. Cited in Department of Energy, Mines and Resources (1976, p. 41).

18. See appendix to this chapter.

Appendix 3A: Organizational Structure—A Historical Perspective

A historical account of the interplay of market forces, technological developments, organizational changes, and governmental and private interference with the working of the free market might contribute to our understanding of the present structure of the world zinc industry.[1] Until the beginning of the nineteenth century, the zinc industry, like all other nonferrous metal industries, was very little developed beyond the use of the alloy form for ornaments and some household wares. Industrial development during the nineteenth century and some developments in the science of metallurgy in the first half of the twentieth century increased the production and consumption of zinc enormously. The intermittent recession years witnessed the development of some formal cartels in the European countries, combines in the United States, and various protectionist policies in several parts of the world.

In the post-world-war period, after a temporary halt, the industry was again revived by the Korean War boom. In the subsequent recessionary period, while the excess capacity in the U.S. industry was protected by a number of governmental measures such as the accelerated stockpile program, quotas, and some incentive programs, the rest of the world had to observe a cutback in production levels. Better business conditions and a surge in demand due to the Vietnam War in the early sixties resulted in a substantial revival of the world zinc industry. In the period 1963–1964 alone the price of zinc doubled. On the other hand, continuously larger-scale production of aluminum and plastics increased their capability as substitutes for zinc in some of its major end-uses. This was almost immediately recognised by the major zinc producers and induced them to agree upon a fixed-price system which could be manipulated by them according to the current circumstances, rather than depending on the very unstable free market. To gain insight into these developments, the discussion will be divided into (1) the pre-world-War period, and (2) the interwar period. Post-world-war II developments are discussed in detail in chapter 2 and hence will not be elaborated here.

The Pre-World War I Period

Earlier in its history zinc was most well-known as an alloying material with lead and copper. The oldest-known piece of zinc (containing 87.5 percent zinc, 11.5 percent lead, and 1 percent iron) was found in the form of an idol in the prehistoric Decian settlement at Deroseli, Transylvania. Brass making seems to have been known to Romans and Asians in the pre-Christian era; some Roman coins as early as 200 B.C. contain an intentional addition of brass secured by melting copper with calamine (the basic mineral containing large quantities of zinc). Between the sixteenth and eighteenth centuries, Portuguese navigators brought zinc to Europe from India and China, where metallurgical science was perhaps more developed than anywhere else in the world. Until the beginning of the nineteenth century, all the requirements of zinc in Europe were satisfied through imports from these two countries.

Calamine ores were first distilled at Bristol, England and were later transferred to Silesia in 1798, and to the United States in 1835. In fact, the basis of the modern zinc industry may be said to have started with the Abbé Dony zinc smelting at the Liège and Vieille Montagne company in Belgium in the early nineteenth century. During the nineteenth century the smelting industry remained largely in Belgium and Germany, which together accounted for about 70 percent of the world production in that century. At the beginning of the twentieth century the United States outstripped Belgium, and by 1909 had also surpassed Germany, becoming the world's largest producer. These three countries together accounted for about 80 percent of the world output of zinc metal, though only 55 percent of the world mine production.

This concentration of metal production was due to several factors, including the delicate and to some extent well-maintained secrecy regarding the nature of the smelting process. The smelting process required highly skilled labor, cheap fuel, and suitable retort clays. The importance of cheap fuel can scarcely be overemphasized, as two tons of coal were required for smelting each ton of ore, thus making it necessary for the ore to move to the sources of coal. It was this technological fact that gave some market power to Belgian and German producers in the world zinc industry, although the major mine production lay in Australia, Spain, Italy, and Mexico at that time. In 1885 in fact, a cartel (the International Zinc Syndicate) was formed under the leadership of Belgian and German producers with an agreement on production quotas.

The outbreak of World War I, fought in the very midst of the concentrated zinc smelting areas of Belgium, Northern France, and Russian Poland, shattered the existing organization. The German and Austrian industries were cut off from the rest of the world. Thus the war smashed a major part of the world zinc metal industry outside the United States, while demand, particularly for high-grade zinc for brass cartridges and shells, was simultaneously increasing very rapidly. An acute shortage of zinc developed, the main bottleneck being the smelting capacity. Vast quantities of zinc ore were lying in Broken Hill lead tailings in Australia, but the European outlets for them were closed. The British government took some active interest in alleviating the problems of the Australian mine producers through a long-term contract to purchase ores, later to be resold to the British, French, and Belgian smelters. The agreement carried a guaranteed price to the mine producers. However, the wartime shipping blockade, and conservatism on the part of the British smelting company in expanding smelting capacity, resulted in the heavy loss of over one million pounds sterling to the British Board of Trade, and an accumulation of over 400,000 tons of zinc ore by the end of the war, rising to 750,000 tons by the end of 1921.

The only country able to take advantage of the wartime increase in demand was the United States, with ample ore, fuel, and skilled labor. The smelting margin rose from $10 per ton in 1914 to $100 per ton by June 1915. As a result, the smelting capacity in the United States doubled within two years (1914–1916) and reached over 900,000 tons by 1917—about three times the domestic requirement of the country in that year, and about 82 percent of the total prewar world requirement for the metal. In the long run this tremendous increase in the smelting capacity could only be justified if much of the former European smelting business could be retained permanently in the United States. This was less likely, as average productivity of the U.S. and Belgian workers was hardly different, while wages were slightly higher in the United States. Further, the failure of the U.S. producers to secure contracts from Australian mine producers rendered more than one-third of the wartime U.S. smelting capacity excess.[2]

By the end of World War I, then, large stocks of concentrates in Australia were building up, together with uncertainty regarding British policy and a large excess smelting capacity in the United States. A European smelting revival subsequently forced the American metal producers to fall back upon the domestic, though much inferior, resources and high tariff walls.

The Interwar Period

The interwar period witnessed many institutional and technological changes leading to some significant alterations in the structure of the world zinc industry.

By 1923 the Western European smelting industry had resumed its operations again. The large Australian stocks had dwindled. The two large Western combines, the Vieille Montagne group and the Anglo-Australian group, dominated the world market outside the United States, though the relations between the chief producers remained highly competitive as each was straining to consolidate his own position. In the United States as well, the industry was organizing more closely under the auspices of the American Zinc Institute and the Zinc Export Association. American interest,through

Anaconda, entered the disputed upper Silesian field and, consequently, the European market.

The period 1923–1928 witnessed the development of the floatation method, a more efficient technique of concentrating ores, which permitted the extraction of zinc from complex sulphide ores, thus augmenting the supply of concentrates enormously. As the ore supply increased, there was again a high premium on smelting capacity, and as a result several mining companies began to build their own smelting plants. A simultaneous development of the electrolytic technique of smelting, however, encouraged a balance of smelting capacity in favor of countries with a cheaper source of energy; in this case, hydroelectric power. Canada, one of the largest beneficiaries of floatation and electrolytic developments, more than doubled her ore/metal production between 1925 and 1928. The Consolidated Mining and Smelting Company of Canada rapidly approached the size of the Vieille Montagne in output and was marketing most of its product in Europe. Some increase in capacity was observed in Australia, Mexico, Rhodesia, and Indochina as well as in Europe. The Silesian-American corporation in Poland, Giesche (a subsidiary of Anaconda of the United States) was campaigning actively for new markets. The United Kingdom was again expanding smelting capacity, and new electrolytic plants were projected in Germany, Norway, and Southern France. Thus, the market power of the earlier European companies was threatened, resulting in a demand for cartelization.

A European zinc cartel, with Vieille Montagne as its leading force, was formed in May 1928. All the important European output was represented, including the Belgian Vieille Montagne, the French branch of Vieille Montagne, the Union des Usines à Zinc, the Dutch Zincs de la Campine, Austurienne, Penarroya, the Polish Oberschlesische Zinc, Giesche (Anaconda Company), Hoheulohe, the German cartel, the English National Smelting Company and Sulphide Corp., the Norwegian Zinc Co., and the Spanish Austurienne and Penarroya. American producers were also represented in order to discuss conditions under which the United States could join the scheme. The cartel, however, went through many extensions and partial and total extinctions until 1935, when it was finally dissolved.

Initially, the agreement was only for six months and, during that period, producers did nothing more than organize informal discussions on the problems. No agreement could be made on price stabilizations, stocks, or production controls. The LME price, after a temporary stiffening in May, again slid off, and by October had reached a new low point.

The first agreement on production control was arrived at in January 1929, when the members agreed to curtail output by 7 percent until the LME price was stabilized at £27 per ton for at least a month. The European electrolytic production was not included in this output restriction scheme. Australian and Canadian producers, although not members of the cartel, were understood to have agreed to restrict their exports to Europe by an equal amount.

The control on production varied between 5 to 10 percent over the first six months, finally stabilizing at 10 percent at the end of the year (1929). It did not take long for friction to show up between different cartel members. The major objection came from the custom smelters who apparently did not secure the same advantage from the cartel as was available to the integrated smelters with mines. The advantage for custom smelters lay in their operation at as high a level as possible regardless of the price, whereas the ore producers and integrated combines would support restriction if it brought about a compensating increase in price. Another weakness of the cartel was its failure to include the growing electrolytic production of the new world, particularly Canada, Australia, Rhodesia, and Mexico. As a result, the cartel was dissolved at the end of 1929.

Several other attempts were made to revive the cartel under Belgian leadership, but these could not succeed because of the various conflicting interests. The major stumbling block was the conflict of interests between the European smelters (where smelting capacity was mainly based on the traditional retort process) and the new electrolytic producers elsewhere. The clash of interests arose because of the nature of marginal cost curves in the two processes. The electrolytic costs are largely for power, while the costs in the retort process are for labor, fuel, and miscellaneous supplies. Under normal business conditions both processes have similar costs, so that choice depends on locality and ore. Under abnormally depressing business conditions, however, operating costs of the retort process decline with lower prices. Under these circumstances, the operating costs of the electrolytic process (where more than 90 percent of the cost of the hydroelectrolytic plant consists of interest and other capital charges on the original investment) remain stable and increase with per-unit decrease in production. Electrolytic producers were consequently more reluctant than retort smelter producers to curtail output. In fact, electrolytic producers increased their smelter capacity by about 50 percent during 1930.

As a consequence of these expansions in smelter capacity, stocks increased by about 136,000 tons in a year and the LME price dropped from the 1929 average of £24.8 per ton (as against £36.6 in 1925) to £13.8 per ton in December 1930. By May 1931, price had further declined to £10.5 per ton and stocks had increased by another 57,000 tons. At this stage, even the electrolytic producers were alarmed.

In July 1931, therefore, a world cartel was formed at Ostend where Belgian, German, Polish, Norwegian, French, Czech, British, Mexican, Dutch, Spanish, Italian, Australian, and Rhodesian producers signed the agreement. The cartel was planned to exist for five years beginning 1 August 1931, but could be dissolved any time at three months' notice. Production capacities were prorated on the basis of the highest three months' output between January 1927 and June 1930, with special allowances made for the new plants of the Hudson Bay Mining and Smelting Company in Canada and the Royal Austurienne in Norway.

The initial restriction was drastic—45 percent of the theoretical capacity. One-half of the existing stocks were permitted to be sold in addition to the current production, the balance to be held for higher prices later. Stocks of 227,000 tons in July 1931 had been cut to 209,000 tons by the end of the year, and 88,000 tons were frozen by agreement out of the later stock.

Within a year, the depreciation of sterling created new problems for the continuation of the cartel. Currencies fluctuated throughout the world; Australia, with her currency depreciated by 50 percent, was eager to sell zinc and could get a good profit by selling at the world price. Nationalistic policies of England and Germany, through imposition of tariffs, further worsened the situation.

By the end of 1932 a new dissension, mainly over the question of stocks, developed within the ranks, resulting in the dissolution of the cartel for about three months, after which a new agreement was reached. Producers with large stocks were eager to liquidate part of them, while producers with no stocks to be held off the market desired that the production restrictions be eased. To solve the deadlock, a system of penalties for overproduction and bonuses for underproduction was introduced, by which the lowest-cost producers could restrict production further. With another agreement on frozen stocks in March 1933, the cartel continued, albeit struggling.

Producers, mainly electrolytic, continued to produce in excess of quotas and to pay fines. In spite of large excess capacity, new expansions continued. Nationalistic policies in Italy, Germany, and some other countries expanded smelter capacity to attain their objective of self-sufficiency. The political revival of silver was another factor for continued expansion, as silver is largely available as a coproduct of zinc.

Thus, the problems posed by stocks, fluctuating exchanges, tariffs, continued capacity expansions, and above all, the heterogeneity of interests of the cartel members, worked for its final dissolution in 1935. The cartel's major weakness, which contributed most to the heterogeneity of interests, lay in the control of smelter production rather than control of both smelter and mine production. In general, absence of control over mine production simply resulted in the accumulation of large stocks of concentrates rather than large stocks of metal. In fact, these concentrates, given the excess smelter capacity, could be rapidly converted into metal and thus threaten the purpose of the cartel at any time.

Such was the history of the zinc cartel, probably much like attempted cartels in many other industries. The institution underwent continuous struggle from the time of its very inception to its final dissolution. It is clear that if there is no other market distortion, under normal economic conditions cartelization in an industry may breed inefficiencies in the market. However, this statement needs to be qualified.

First, according to the theory of the "Second-Best," an addition of one imperfection, say cartelization, in the presence of even one other imperfection in the economy, need not necessarily result in a loss of efficiency. Besides innumerable imperfections in the economy as a whole, the zinc industry, probably like many other mineral industries, has been subject to government interference through tariffs, quotas, stockpile programs, national monopolies, and so forth. In this case, one cannot conclude that cartelization must have resulted in the loss of efficiency in the world zinc industry. Second, according to a pragmatic view, though cartelization may result in a check in technical advances under normal business conditions through the protection of inefficient units in conditions of deep depression, like those existing at the time of cartelization in the zinc industry, it works as a medicine for the patient units, while a laissez-faire policy works for their untimely death.

During the 1929–1934 period, the main problem faced by the industry was the temporary reduction in demand due to a general trade depression. Under laissez-faire conditions, the higher-cost concerns would more-or-less have been ruined before they decided to close down; they might not have been able to stand the drain of maintenance costs, with the result that when demand recovered they could not in time resume efficient production. Thus a wholly unnecessary boom would have been generated to induce the establishment of new concerns to take their place. This could have been a completely unnecessary loss to the shareholders of these concerns and could have meant a completely unnecessary absorption of new capital from the point of view of the community. Restriction schemes can effectively prevent these unnecessary and wasteful results of laissez-faire. The need in this case "is to prevent the extinction of capacity, or in other words, to put that portion of the existing capacity which is temporarily unwanted in the cold storage, so that it may be preserved in a fresh and efficient condition against the day when the depression passes, and it will be again required; and, one should add, to accomplish this at the minimum cost. Thus restriction schemes are an excellent form of refrigerator, and reliance can be placed upon them, at least in theory, to enable an industry to survive a severe depression without demoralisation and decay."[3]

However, this does not necessarily imply that restriction schemes are desirable whenever there is excess capacity in the industry. On the contrary, when general trade conditions are normal, protection of excess capacity will only breed inefficiency. Protection of higher-cost producers through cartelization can only postpone the evil for a later day. In the event where technological progress has induced new capacity, making some of the old capacity excess, restriction of such an excess capacity will only hold the technological advancement back behind the restriction scheme; and such schemes can hardly continue for long as new additions to capacity will keep reducing prices. The technically obsolete capacity must be surrendered to laissez-faire to perform the necessary surgical operations. And, in fact, some of it may be economically obsolete and hence need to be scrapped.[4]

However, at least three qualifications to this general observation may be noted. In the first place, if the excess capacity is accompanied by a temporary fall in demand due to a world trade depression, restriction schemes will be justified, as argued above, only for the depression period. It is true that the consumer is likely to bear a considerable part of the cost of prolonging the life of an economically obsolete capacity; that is, the burden of payments to the owners of technically but not economically obsolete capacity in order to preserve their existence. But these losses may be weighed against the general economic and social disturbances which accompany the surgical operation of laissez-faire. Second, protection of excess capacity may also be justified under conditions of normal business but technological stagnation: if technique is stationary, the excess capacity, by hypothesis, is technically efficient and may therefore be refrigerated until it is required again. However, in this case, since the additional capacity created must be due to the mistakes of entrepreneurs in judging the growth in demand, the cost of cold-storage must be borne by the entrepreneurs rather than the consumers; that is, there should be no increase beyond normal prices under restriction schemes. Finally, if the excess capacity appears in a few countries as a result of governmental interference with the market in other countries (for example, inducements through protective measures), restrictions in the former countries may not be totally unjustified.

Notes

1. The discussion in this appendix is based on various fragmentary evidences in numerous publications on zinc. For overall completeness, the interested reader is referred to Elliot et al. (1937, chapters 2 and 12) and McMahon et al. (1974).

2. Although Mexican and Canadian supplies of ore to the United States were increasing, it was important to secure contracts from the Australian producers, as they supplied more than one-third of the world production of zinc ore.

3. See Rowe in Elliot et al. (1937, p. 79).

4. In the sense that its prime costs exceeded the total costs of the newest capacity.

4 Econometric Modeling of Mineral Industries: A Survey and Specification of a Model for Zinc

Commodity modeling in the last decade has emerged in a number of analytical forms, according to the objective of the researcher and the particular behavior of the decision maker to be modeled.[1] Econometric process models designed to analyze industry processes, world trade models to study transmission of short-run fluctuations of domestic activities, and systems models that facilitate the study of behavior patterns of decision makers in reaching equilibrium or adjusting to various constraints, are among the recent developments in commodity model building. More recently, attempts have also been made towards incorporating (1) market imperfections, (2) technological considerations in the modeling of certain commodities, and (3) the linking of commodity models to macroeconometric models of the important consumer and producer countries.[2] Since the general field of commodity modeling has been very well surveyed and analyzed recently, focus here will be restricted to the techniques followed by some model builders concerned with mineral commodities.[3]

Models of Mineral Commodities: A Survey

Modeling of mineral commodities, except for oil, has been based in general on the technique termed "econometric market modeling."[4] The technique consists basically of laying down a set of market relationships pertaining to the supply of and demand for a commodity, together with inventory behavior, and their roles in determining the price of the commodity. Prices, along with some exogenous variables, affect the supply, demand, and stock variables which, in turn, determine the equilibrium level of price and quantity of the commodity. Various technological and institutional variables relevant to the particular industry, or an emphasis on particular market forces, or on the behavior of decision makers, distinguish these models from one another. These models have the advantage of being easily amenable to microanalysis of the market, for example, to stabilization schemes through simulation techniques, and to macropolicy analysis through their linkage to the macroeconometric models of producer or consumer countries.

A general scheme of the market form of an econometric model for a mineral commodity may be depicted through an arrow diagram as in figure 4–1.

Many variations of the following scheme are possible, depending on the objective of the model builder, the particular type of behavior of the decision maker to be emphasized, and other relevant considerations with regard to the particular commodity. For example, it may be important to build a model ignoring, or paying very little attention to one or more variables, such as resources, capacity, or some technological/institutional aspects. Or, it may be required to ignore one of the major market variables such as supply, or demand, or to link some of the market variables to relevant macroeconomic variables, as warranted in the particular situation. Here, a brief review of some of the models of mineral commodities is necessary to illustrate the techniques of model building followed in this field.

Tin

The model of the world tin market built by Desai (1966) follows the scheme described, with some important variations. Desai's major objective was to study the transmission of fluctuations from the developed world to the developing countries. This model, therefore, had a very simple structure of a recursive nature:[5]

$$D_t = D(A_t)$$

$$S_t = S(S_{t-1})$$

$$\Delta STK_t = S_t - D_t$$

$$P_t = P\left(\frac{STK}{D_{t-1}}\right) \tag{4.1}$$

The symbols are defined in figure 4–1; the subscript t stands for time.

The model was disaggregated on the demand side into three regions: the United States, OEEC and Canada, and the rest of the world. The total demand for tin in the former two regions was further disaggregated according to two end-use categories—tinplate and non-tinplate—to capture more accurately the influence of the relevant activity variables and technological changes in the end-uses. The immediately relevant variables relating to the use of tin for tinplate and non-tinplate were linked with larger macro-variables, such as GNP and industrial production. Price variables did not contribute to the explanatory power of either the supply or demand functions, and hence were excluded.

The tin model was used to study the transmission of cyclical fluctuations in the activity variables of the industrialized countries to the prices and total

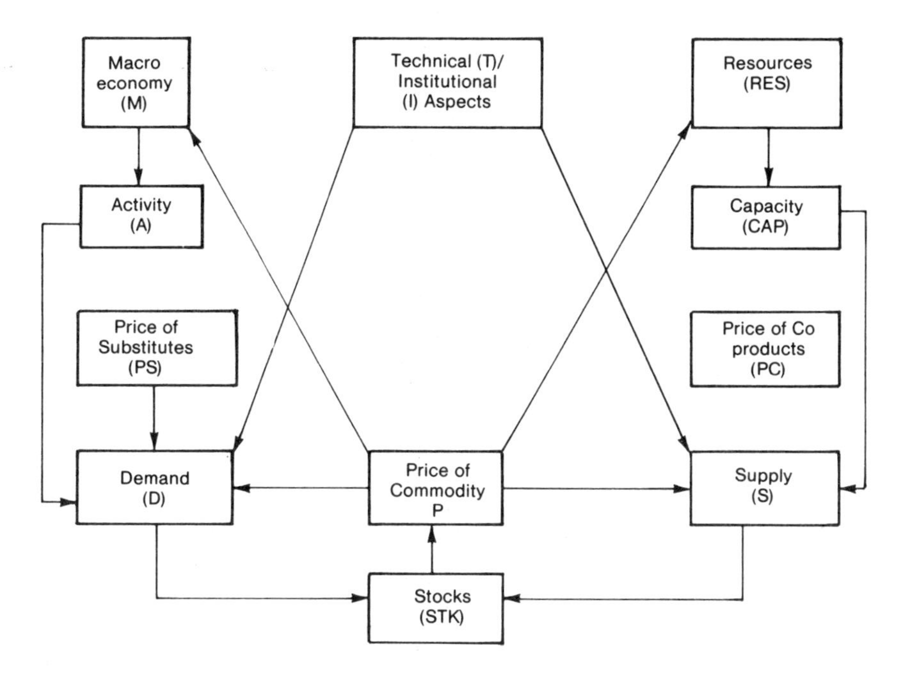

Figure 4–1. Flow Chart of Market Econometric Models (Mineral Commodities)

revenues received by the major suppliers of the tin market. The technique of stochastic simulation was used for this purpose.[6] Other simulations carried out were aimed at investigating the possibility of reducing fluctuations in price and revenue received by tin producers, through the instruments of a buffer stock and the restriction of output by the International Tin Council.

Copper

Many researchers have attempted to build a model of one or more phases of the world copper market.[7] Fisher, Cootner, and Baily (1972), however, presented the most comprehensive study of the world copper industry. This study follows essentially the market form of econometric modeling technique as discussed above. Their model may be written as:

$$S_t = S(P_t, P_{t-s})$$

$$D_t = D(P_t, P_{t-s'}, A_t, PS_t)$$

$$P_t = P\left[\Delta\left(\frac{STK}{D}\right)_t, P_{t-1}\right]$$

$$\Delta STK_t = S_t - D_t \qquad (4.2)$$

Variables are as defined in figure 4–1. P_{t-s} indicates distributed lag response (in terms of partial adjustment model) of S and D to prices. The subscript s indicates the number of years lagged.

The copper model divided the world copper market into the United States, where prices are administered by the U.S. government and U.S. producers, and the rest of the world, where prices are determined by free market forces of demand and supply at the London Metal Exchange (LME). Since the LME price is a free-market price, it also plays a role in determining the U.S. producer price in the long run, as well as providing a link between the two markets. Interregional trade between the United States and the rest of the free market world, that depends on the differential between the two market prices, provides a further link between the two markets. The model was relatively disaggregated by incorporating different supply equations for the major copper-producing areas (United States, Chile, Canada, Zambia, and the rest of the world) and different demand equations for each of the principal consumer areas (United States, Europe, Japan, and the rest of the world). The demand equations were, however, not disaggregated to end-use categories. Neither were resources, capacity, technological variables, and prices of coproducts included in the model. Nevertheless, the model remains one of the best examples of the market form of econometric modeling.

The estimated version of the model was used for short-term forecasting (though not very successfully) and policy-simulation analysis. The major policy questions asked include (1) the effect of a 10-percent rise in Chilean output every year on price level and on Chilean revenues, and (2) the effect of a discovery of a large new source of supply on the LME price.

Cobalt

Burrows's (1971) model of cobalt introduces market imperfections explicitly. Unlike copper, tin, and many other mineral commodities, production of cobalt is highly concentrated. One company, Union Minière Haut Katanga (UMHK) produces more than 60 percent of the world output, the rest being produced by various companies in Canada (8 percent) and many other countries. Such a concentration on the supply side rightly warrants allowance for market imperfections in model specification. The general structure of the model is derived by treating UMHK as a price setter following profit-maximization principles (given the supply response of all other producers).

Profit maximization, given the world demand for cobalt and the supply response of the other producers at the prices set by UMHK, yields the price-determination equation for cobalt. Although the consumption structure of cobalt is fairly detailed according to end-uses in the United States, the model lacks determination of the rest of the world's (ROW) cobalt consumption and UMHK production behavior, which were later included by Adams (1972). The U.S. government's General Services Administration stockpiles (GSA) are explicitly introduced in the price equation, these being looked upon as potential sources of supply by the producers of cobalt.

The general structure of Burrows's cobalt model may be represented as follows:

$$D_t^{USA} = D(P_t, \bar{P}_{t-s}, \bar{A}_t)$$

$$D_t^{ROW} = \bar{D}_t^{ROW} \; or \; D_t^{ROW} = D(P_t \; \bar{P}_{t-s'}, \bar{A}_t) \text{ Adams (1972)}$$

$$S_t^{UMHK} = D_t^{USA} + \bar{D}_t^{ROW} + \overline{\Delta GSA_t} - \bar{S}_t^{ROW}$$

$$S_t^{ROW} = \bar{S}_t^{ROW} \; or \; S_t^{ROW} = f(\overline{Time}) \text{: Adams (1972)}$$

$$P_t = P(D_t, \overline{GSA_t}, \overline{\Delta GSA_t}) \tag{4.3}$$

Notations are as explained in figure 4–1 and in the preceding paragraph; a bar over a symbol indicates that the variable was not endogenously determined by the model.

Zinc

One of the most important nonferrous metals, zinc, seems to have been neglected by the commodity-model builders. One may speculate on the reasons for this neglect, however. Although zinc is very important for the manufacture of many durable commodities,[8] neglect of this material for model-building may be attributed to (1) the relatively small cost of zinc in the total cost of most final commodities which use it; (2) the fact that, except for Mexico and Peru, all other major producers are developed countries where foreign exchange earnings from zinc are relatively less important than they would be in developing countries, and (3) the fact that some efforts at UNCTAD to study the world zinc industry were left incomplete.[9]

Thus, the present study, to the author's best knowledge, is the first attempt to carry out a comprehensive analysis of the world zinc industry based on a market form of econometric modeling, as outlined above.

Organizational Structure and Modeling of the World Zinc Industry

The investigations into the organizational structure of the world zinc industry in the last two chapters reveal that about 52 percent of the mine output in 1974 was localized in four countries: Australia, Canada, Mexico, and Peru. Including their multinational operations, twenty-four corporate groups had controlling interests in about 65 percent of the free-world mine capacity. About eleven companies (having a mining capacity of more than 100,000 tons, including their multinational operations), shared control of about 55 percent of the FME world mine capacity in 1974. In the same year, the 7 (4) largest companies, having a mining capacity of more than 200,000 tons (4 percent of the FME world mine capacity, including the multinational operations of the companies), had controlling interests in about 43 (32) percent of the FME world mine capacity. Under these circumstances it is very unlikely that the producers will be successful in achieving any formal or informal collusion in the industry.[10] Lessons from such attempts during the interwar period, when the industry was even more concentrated, support this proposition.[11] Further, there are at least two reasons why it may be reasonable to assume competitive behavior in the world zinc industry.

In the first place, Stiglitz (1976) has shown that, in general, there is very little scope to exploit monopoly power in the extractive resource industries. In fact, under the assumptions of constant elasticity-of-demand schedules and zero extraction costs, monopoly price and competitive price are identical. In some other cases, a monopolist is more conservation-minded than a competitive firm.

The basic argument is very simple. In a two-period model with a constant elasticity-of-demand schedule and zero extraction costs, a competitive producer must be indifferent to selling the last unit of exhaustible resource in period t or $t+1$, so that the market equilibrium is the point of intersection of the two demand curves D_t and D_{t+1} (figure 4–2). The monopolist, on the other hand, compares the marginal revenue in period t (MR_t) with the discounted marginal revenue in period $t+1$:

$$\left(\frac{MR_{t+1}}{1 + r}\right) \tag{4.4}$$

where r is the rate of discount. With the assumption of constant elasticity of demand schedules in both periods, which implies that marginal revenue is proportional to price, the two equilibria (competitive and monopolistic) yield the same level of extraction, $Q_t = Q^*$; and prices,[12]

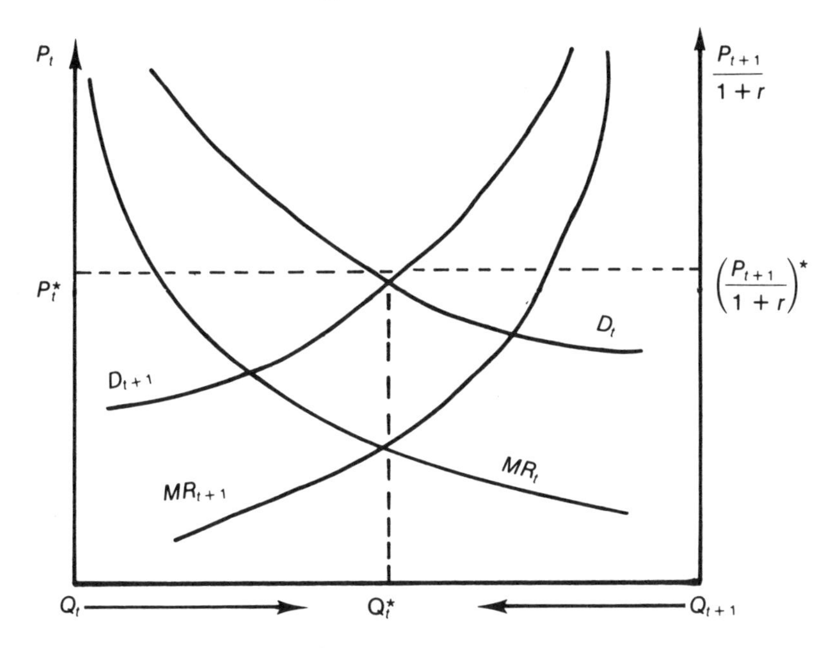

Figure 4–2. Exploitation of Mineral Resources under Competition and Monopoly

$$P_t^* = \left(\frac{P_{t+1}}{1 + r}\right)^*. \tag{4.5}$$

If the elasticity of demand in year $t+1$ is higher than in year t (a larger possibility of substitution in the long run), the ratio of price to MR will be higher in $t+1$ than in t, which means, at the competitive price:

$$\frac{MR_{t+1}}{1 + r} > MR_t, \tag{4.6}$$

inducing the monopolist to sell more in the next period. The monopolist is more conservation-minded than the competitive producer. The same result holds for nonzero extraction costs with constant elasticity of demand. However, for extraction costs rising with the extraction of resource (that is, costing more as less mineral is left in the ground), the result is not clear. Further, the rate of discount may change; variations in demand and costs in the future may be too uncertain to be predictable; and, a monopolist conserving the resource may run the risk that a large new source of supply or a very cheap substitutable material will be discovered. All of these cases need further analysis before we can be sure about the identity of monopolistic and competitive equilibrium prices for exhaustible resources. However, these results do provide an indication that the scope for a monopolist to exploit his market power in the exhaustible-resource industry may be rather limited.[13]

Second, possession of market power does not guarantee that the market power will be used. Quite often, in fact, concentration in an industry may contribute toward the realization of the theoretical results of pure competition through provision of otherwise-unavailable information to the market participants, and hence may save the industry from recurrent short-run fluctuations due to the operation of an invisible hand. Something of this nature seems to be present in the world zinc industry. As noted in chapter 3, major producers and consumers in the world zinc industry do associate themselves under ILZSG, which gathers information for its members on likely changes in market demand and supply to avoid recurrent short-term flucuations in the market.

In the light of these discussions, this model will be based on a competitive-industry hypothesis. However, the U.S. market will be treated separately, since the existence of restrictive practices as carried out by both producers and government in that country, is established.

A General Specification of the Model of the World Zinc Industry

To study the structure, behavior, and performance of the world zinc industry in a generally competitive framework, and to investigate the implications of certain policies for the future of the world zinc market, a market form of the econometric model seems to be the most appropriate one. In this section a general outline of the model will be given, leaving details of specification and econometric estimation for the next chapter.

The noncommunist world zinc market has been divided into two parts: the market administered in the United States and the free market in the rest of the world, for the reasons discussed above. The markets are linked through prices and exchange rates.

The complete model, as outlined in figure 4–1, includes equations for demand, supply, prices, and stocks. Interrelations of these variables in an aggregative and a simplified version of the model are depicted in figure 4–3 for convenient reference.

Supply of Zinc

The supply behavior of zinc producers is distinguished as between primary producers and secondary producers (those who recover zinc from scrap).

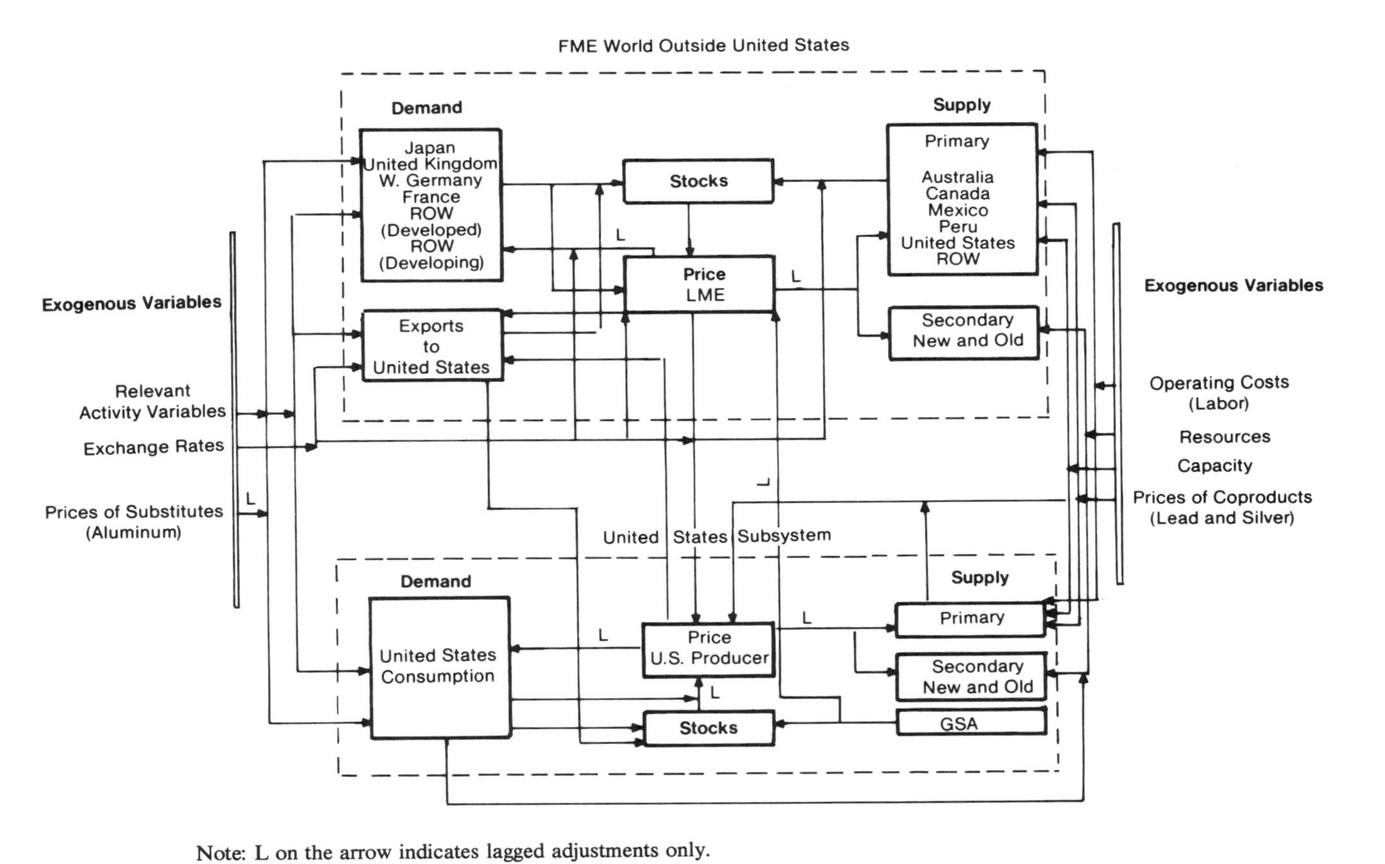

Note: L on the arrow indicates lagged adjustments only.

Figure 4–3. A Model of the World Zinc Market: An Aggregative and Simplified Version

Economics of Mineral Resources and Primary Supply of Zinc. Given the discovery and development of a mineral deposit, the mineral substance becomes a stock. A decision regarding the time-path of extraction from the stock depends on the present as well as the expected future economic environment, and hence can be regarded as a problem in dynamic optimization.[14] Here, an outline of the basic ideas that may throw some light on the inherent difficulties in the modeling of the supply side will be given.[15]

Given the stock of resources K, producers can choose to supply more in the future and less in the current year, or vice versa. The most profitable behavior involves maximizing the net present value NPV of the sum of the future and current revenues Π until the stocks K are exhausted. The problem then is

$$\text{Max}\, NPV = \int_{t1}^{t2} \Pi[q(t),t]e^{-rt}dt - \lambda \left[\int_{t1}^{t2} q(t)dt - K \right]$$

$$\text{subject to} \int_{t1}^{t2} q(t)dt = K \tag{4.7}$$

where $q(t)$ is the rate of extraction at any time t, r is the continuous discount rate, λ is a Langrangian multiplier, and $t1$ and $t2$ the initial and the terminal dates.

Solution through the calculus of variations yields[16]

$$MR(t) = MC(t) + \lambda\, e^{rt} \tag{4.8}$$

where $MR(t)$ and $MC(t)$ are marginal revenue and marginal costs at time t, respectively.

This is a familiar condition for the equilibrium of the firm except for the term $\lambda\, e^{rt}$, which represents a user cost—a sacrifice of future revenue because of sales in the current year, and is a constant.

The formulations become much more complicated if allowances are made for dependence of cost on the amount of total output extracted, or relaxation of assumptions regarding perfect certainty about future prices of output and inputs, constant rate of discount, constant stock of resources over time, and so forth. Further, in such cases theory loses much of its empirical significance.[17]

While keeping in mind the implications of the theory of exhaustible resources, in this study the derivation of supply schedules will be based on the assumption of current profit maximization; an assumption which has, in many earlier studies, provided a good approximation in the analysis of mineral industries.[18] Thus, the following supply function for mine output is assumed:

$$MP_t = f_1 (PZ_{t-1}, W_t, CAP_t, T, PC_t) \quad (4.9)$$

where MP is mine production of zinc; PZ_{t-1} is the price of zinc, lagged one year; W_t is an index of variable factor prices; T is an index of technological change; PC_t represents the prices of coproducts (lead and silver, particularly); and CAP_t is mine capacity.

Production, however, may not respond instantaneously, because of various lags involved in the process of adjustment to variation in prices. In general, it takes a long time to explore new resources or increase mine capacity. Exploration activities require huge investments, which are quite risky since the efforts may turn out to be unsuccessful. For this reason, exploration companies do not undertake new ventures until they are convinced of a long-term rising trend in prices. This may itself involve more than five to six years, in addition to another five to six years required for successful exploration and for making deposits suitable for exploitation. A change in plant capacity in response to price variations may also involve long lags, for the expansion of capacity is quite capital-intensive, and once expansion has taken place it may be very costly to close an operation if the price rise turns out to have been only temporary. Usually, in recessionary situations, mine producers have to continue to operate at the current capacity levels, even though they may be making losses.[19]

In the case of supply, therefore, it will be assumed that only a partial response is available in any particular year. For ease of exposition, suppose there is a simple supply function

$$MP^*_t = \alpha + \beta\, PZ_{t-1} + u_t \quad (4.10)$$

where * indicates desired quantity; PZ is the price of zinc; and u_t is a random error component. A partial adjustment process is assumed as defined by

$$MP_t - MP_{t-1} = \lambda\, (MP^*_t - MP_{t-1}) \quad (4.11)$$

That is, the change in supply in the current year is equal to a fraction λ of the difference between the desired level of supply MP^*_t and last year's actual level of production MP_{t-1}.

Substituting 4.9 into 4.11 and rearranging terms, we have

$$MP_t = \lambda\alpha + \lambda\beta\, PZ_{t-1} + (1 - \lambda)\, MP_{t-1} + \lambda u_t \quad (4.12)$$

where $\lambda\beta$ is the short-run effect of price on supply, and β measures the longer-run response to price. Given the lags involved in exploration, capacity expansion, and implementation, the producers' response to variations in prices is expected to accord with this adjustment process.

Secondary Supply. Secondary supply SCRAP may be divided into new scrap *NS* and old scrap *OS*. The supply of new scrap, which is generated in the process of fabrication of the final product, may be assumed to depend on the level of consumption of the metal *CN* and the price of zinc *PZ*. Thus,

$$NS_t = f_2\,(PZ_t,\ CN_t) \tag{4.13}$$

However, metal recovery from old scrap, the discarded final products which contain zinc (such as automobile scrap), is more involved. Often, piles of old scrap that have accumulated over time are termed "surface mines." Recovery of metal from these surface mines has to compete with the primary resources available. Minerals available in deposits in larger quantities will discourage the exploitation of surface mines in the same way that higher-grade deposits get a priority (due to lower costs) over lower-grade deposits. In general, given the amount of primary resources, a rise in consumption level may be assumed to attract one's attention towards secondary resources, particularly old scrap. Further advances in the technology of metal recovery from scrap which reduces cost of recovery, or a rise in the price of zinc, in general, will also increase recovery of metal from scrap.

Thus, the supply function of zinc recovered from old scrap will take the form

$$OS_t = f_3\,(PZ_{t-1},\ CNRES_t,\ TIME), \tag{4.14}$$

where *CNRES* is consumption of zinc relative to primary resources and *TIME* is a trend variable used to capture the influence of changes in technology. A one-year lag in the response of production to price is assumed for the reasons given above.

Demand for Zinc

Zinc, an intermediate input, is consumed by various industries such as construction, steel, automobiles, rubber, and many other manufacturing industries. Further, although many of these industries may be imperfectly competitive in their domestic markets, none of them may be able to influence the world zinc market to any significant degree. It is assumed that these consumers of zinc try to minimize the cost of zinc (an assumption not incompatible with many forms of market behavior, such as minimax behavior, Baumol's sales maximization, or some broader classes of satisficing behavior) in making their decisions regarding the use of zinc in their final products.

Demand for zinc *CN* then may be assumed to depend on the price of zinc *PZ*, an activity variable *A*, and the price of substitutes *PS*. Thus,

$$CN_t = f_4\left(\sum_{s=1}^{T} PZ_{t-s}, A_t, \sum_{s=1}^{T} PS_{t-s}\right) \tag{4.15}$$

where T is the terminal date for response of consumption to prices.

As in the case of supply, the response of demand to price changes may be quite slow. A slow speed of adjustment of demand to variation in prices may be attributed in part to the fact that zinc, as an intermediate input, accounts for a very small proportion of the total cost of most final products. For this reason, small variations in the price of zinc in the short run, do not alarm consumers to the extent that they seriously consider replacing it by some other material. This behavior is strengthened by the fact that most of the manufactured products that use zinc would require changes in technological aspects of the producing plant, and this might not be undertaken until the price change had persisted in the same direction for a considerable period of time. This also implies that consumers base their choice on past prices when deciding to install a particular manufacturing technology or process that is suitable for the use of zinc in their final product. In fact, in attempts to estimate demand functions for zinc, it is hard to find current price coefficients to be statistically significant. Quite frequently current price coefficients were found to be wrongly signed. For this reason, it is postulated that current demand depends only on lagged prices.

Various lag structures were considered, the most successful being a polynomial lag structure[20] where price response first increases up to the third, fourth, or fifth year, and then tapers off (inverted V-lag) gradually, depending on the nature of the industry and country using the zinc.[21]

Determination of the Price of Zinc

In this model there are two prices, for the reasons discussed above: the U.S. producers' price *USPZ* and the price in the free-market world outside the United States *LMPZ*.[22] First, free market price will be determined.

The Free Market Price (LMPZ). The free-market price being as it is, must depend on market forces of supply and demand. In general, a rise in supply in relation to demand will depress prices and vice versa. In the case of durable goods, particularly mineral resources, however, this excess demand is reflected in variations in stocks *STK* which therefore play an important role (see figure 4–4) in the determination of prices.[23] In general, in the free market prices will adjust to their normal level until the stocks held by the producers reach a satisfactory level in relation to their sales. A higher stock consumption ratio *STKCN* will induce prices to move downward; conversely, a lower ratio will induce an upward movement.

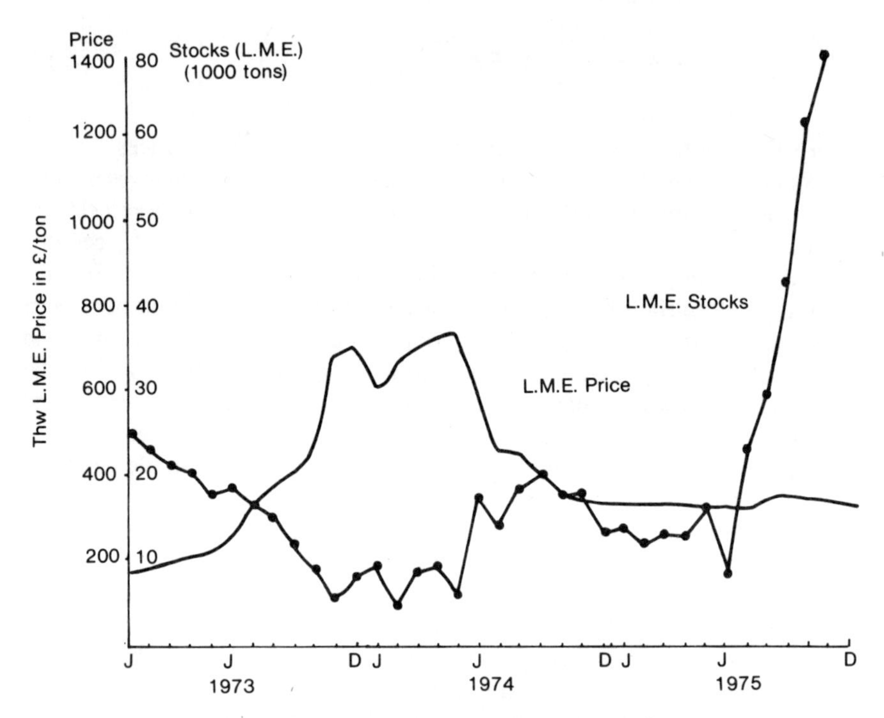

Source: *Yearbook* (New York: American Bureau of Metal Statistics, 1972–1976).

Figure 4–4. The LME Stocks and the Price Behavior of Zinc

Besides, in the world zinc market the stockpiles held by the U.S. government *GSA* have also influenced the price considerably. Often the *GSA* is looked at by the producers of zinc in the free world as a potential source of supply. Higher levels of *GSA* will, therefore, have a depressing effect on price. This has been a recurrent psychological feature of the world zinc industry as reported in many issues of the *Engineering and Mining Journal*.[24] Thus, GSA_{t-1} (*GSA* lagged by one period) and ΔGSA_{t-1} ($\equiv GSA_{t-1} - GSA_{t-2}$, to capture the further lagged effect, if any) will be included in this price-determination equation. As adjustment of prices to stocks may not be instantaneous due to many rigidities in the system, as discussed above, the price of zinc lagged by one period is also included to capture a longer period of adjustment. Thus, the free-market price *LMPZ* is given by

$$LMPZ_t = f_5\,(STKCN_t,\ GSA_{t-1},\ \Delta GSA_{t-1},\ LMPZ_{t-1}) \qquad (4.16)$$

U.S. Producers' Price. LMPZ, being a free-market price, can be looked upon as a long-run equilibrator of supply and demand. All other prices, in this case such as *USPZ*, must show a tendency to converge to the free-market price in the long run. However, in the short run, price is influenced by the U.S. producers and may be affected by varying stocks and the capacity utilization ratio (*USCAPUSE*). In this case, then, it is the change in stocks in relation to consumption ($\Delta USTKCN$) and the capacity utilization ratio that are more meaningful in the determination of the U.S. price.[25] But if *USPZ* drifts too far from *LMPZ* over an extended period, it may become unmaintainable. As a result, U.S. consumers will shift their allegiance from U.S. producers to producers in the rest of the world. Hence, a plausible specification would be

$$USPZ_t = f_6\,(\Delta USTKCN_t, LMPZ_t, USCAPUSE_t). \quad (4.17)$$

Closing the Model

The model is closed by two stock identities (one for the United States and the other for the rest of the world), and an equation relating to net imports from the rest of the world to the United States.

Change in stocks held by the U.S. producers $\Delta USTK$ are equal to U.S. mine production *UMP* plus recovery from scrap $UNS + UOS$ plus net imports into the United States *UIMP* minus U.S. consumption *UCN* and minus increase in U.S. government stocks ΔGSA. Thus,

$$\Delta USTK_t \equiv UMP_t + UNS_t + UOS_t + UIMP_t - UCN_t - \Delta GSA_t \quad (4.18)$$

Similarly, the stock identity for the rest of the world $\Delta RSTK$ is mine production in the rest of the world *RMP* plus recovery of metal from scrap (old and new) in the rest of the world *RSCRAP*, plus net zinc imports from the centrally planned economies *RIMP* minus net exports to the United States *UIMP* minus consumption of zinc in the rest of the world *RCN*.[26] Thus,

$$\Delta RSTK_t \equiv RMP_t + RSCRAP_t + RIMP_t - RCN_t - UIMP_t \quad (4.19)$$

Trade between the United States and the Rest of the World

Trade between the United States and the non-U.S. world, along with the price variable, seems to link the two markets. The trade equation has been specified as an import demand function of the U.S. consumers that depends on the intermarket price differential $LMPZ - USPZ$ and the activity variable in the United States *UA*. Thus,

$$UIMP_t = f_7 [(LMPZ_t - USPZ_t), UA_t] \tag{4.20}$$

The equations 4.12 through 4.20 represent the structure of the model in a highly simplified form. As shown in the next chapter, the models that are used for estimation are fairly detailed and disaggregated. Two versions of this model will be developed. In one version supply and demand equations by the major producer and consumer countries are disaggregated to capture the essential differences in the consumption and production patterns of these countries. In the other, the demand for zinc in all the major consumer countries is disaggregated by the end-uses of zinc to relate the consumption of zinc directly to the relevant user industries. This is important, as industrial structures in different countries vary. The technological aspects of the different uses of zinc and their responses to market variables are more appropriately reflected in this disaggregated version.

From the simplified presentation of the model, it is easy to see that all the equations are identified. Consumption demand depends only on the lagged prices and exogenous variables. Import demand depends on the difference between the two market prices, and so is distinguished from the U.S. consumption demand in general. Producers' supply responds to lagged zinc prices and variable factor prices and capacity variables. Prices depend on the level of stocks in relation to consumption and other relevant variables, whereas identities define changes in stocks.

Another feature of the model, which has important implications for the methods of estimation, is that the system as outlined above is recursive.[27] In the recursive simultaneous structures, use of the ordinary least-squares method of estimation is justified, because the estimators are not subject to problems of inconsistency associated with systems of simultaneous equations in the general case.[28]

Notes

1. For an excellent taxonomy of commodity modeling techniques, see Labys, ed. (1975, chapter 1).

2. For example, models introducing (1) market imperfections, see Epps (1970), Burrows (1971), Dayananda (1977); (2) technological considerations, see Avramidas and Cross (1973); (3) linkage to macroeconometric models, see Adams (1973a).

3. See Labys, ed. (1975); Adams and Behrman (1977).

4. Ibid.

5. A recursive scheme, in simple language, involves a determination of all the endogenous variables in the scheme without any feedback effect; that

is, unidirectional causation. For a technical discussion of recursive systems, see Johnston (1972, p. 369).

6. For the technique of stochastic simulations, see Adelman and Adelman (1959).

7. For example, see Ballmer (1960), Behrman (1972), Mahalingsivam (1969), Khanna (1972), Fisher, Cootner and Baily (1972), Adams (1973), and Banks (1974).

8. See chapter 2 for details. Also, the U.S. Government has ranked zinc as a strategic material, and the United Nations has established a separate study group charged mainly with the collection of statistical material relating to the zinc (and lead) industries.

9. See, for example, Banks (1971). Banks pioneered the econometric study of zinc at UNCTAD, but for some unknown reason did not go further than estimating some demand functions, and left the study of zinc in favor of copper. Another attempt at modeling the world zinc industry that has come to my attention was made by the joint efforts of two private organizations, Charles River Associates, Inc., and Wharton Econometric Associates, Inc. As the study was carried out on contract for the U.S. government, full details of it are not available to the public. Based on what is available, their model was a market form of econometric model, but did not incorporate the supply side. Their objective was to study various scenarios with regard to U.S. government policy relating to its strategic stockpile program for zinc.

10. For example, Scherer (1970, pp. 50–57) has catalogued many industries in terms of four-firms concentration ratios. In general, it is agreed that the existence of interdependence in decision making of the firms (and therefore the oligopolistic behavior) requires the control of at least half of the total output by the four largest firms in the industry.

11. See appendix to chapter 3.

12. In a multiperiod model, with a finite or infinite time horizon, the basic result still holds with the qualification that both price and MR in competitive equilibrium and monopolistic equilibrium must rise at the rate of interest r. For mathematical proof, see Stiglitz (1976).

13. Pindyck (1978), assuming that the cartel is in a position to behave as a perfect monopolist, has computed optimal gains for the three existing cartels in petroleum (Organization of Petroleum Exporting Countries), bauxite (International Bauxite Association), and copper (International Council of Copper Exporting Countries). Whereas, according to this analysis, OPEC and IBA (who account for around two-thirds of non-Communist world production of petroleum and bauxite) stand to gain from cartelization, the same is not true for CIPEC. The zinc industry, then, which is even less concentrated than copper and has no formal or informal cartel at present, does not show any promise for a successful cartelization in the near future.

14. The economics of mineral resources that involve such decision making processes date back to Hotelling (1931), and was developed later by Herfindahl (1955), Gordon (1967), Levhari and Liviaton (1977). A recent comprehensive survey of the literature is by Peterson and Fisher (1977).

15. For mathematical formulations of the problem, the interested reader is referred to Gordon (1967), and Liviatan and Levhari (1977).

16. See Gordon (1967, p. 217).

17. As Gordon himself points out, "Examination of theory's empirical significance suggests several crucial difficulties. The automatic assumption that the pure theory of exhaustion is applicable to natural resources involves a complete misunderstanding." See Gordon (1967, pp. 275–276).

18. For some of these studies, see the first section of this chapter.

19. See the discussion on this point in the appendix to chapter 3.

20. For a detailed discussion of the polynomial distributed lag, see Almon (1965) and Johnston (1972, pp. 292–300).

21. This is quite plausible, as consumers would normally take one to two years to be convinced of the long-term nature of a price rise and would then gradually reduce the consumption of zinc, adapting their plant so as to be suitable for the use of other materials, and later discontinuing use of zinc altogether in favor of other materials. In the process, producers will also have to convince the consumer of their final product about the quality of that product when it is based on some material other than zinc.

22. For a detailed discussion of the price system, see chapter 2.

23. Although in the case of nondurable commodities, stocks may not be important, in the case of commodities such as zinc, stocks play a dominant role in the determination of prices.

24. See also Burrows (1971, p. 154) for discussion of a similar influence of GSA in the cobalt market.

25. For a detailed discussion of the influence of the U.S. producers on the U.S. price, see chapter 2.

26. There was no trade between the United States and the centrally planned economies in zinc during the period to which the model was fitted.

27. Given the nature of disaggregation and specification of the model, it is unlikely for the error terms across the structural equations to be correlated. It is, therefore, assumed that the covariances of the disturbances across the equations are zero.

28. For a discussion of recursive systems and the method of estimation, see Johnston (1972, p. 369 and pp. 377–380).

5 Econometric Estimates of the Models of The World Zinc Industry

Two versions of the model of the world zinc industry, as specified in a general format in the last chapter, were estimated.[1] In the first version of the model, separate demand and supply equations were provided for the major consumer and producer countries in the FME world. On the demand side, as discussed in chapter 2, it is expected that the variations in the demand patterns of the different countries will be captured better in a model which disaggregates the total demand according to the major consumer areas. The most important consumers with some differences in their demand patterns, are the United States, Japan, the United Kingdom, West Germany, France, the rest of the developed world ($R1$), and the rest of the world ($R2$). Similarly, there may also be differences in cost structure, which may be attributed to the nature of existing deposits, and differences in operating costs and in technology in the different countries. Total world supply is, therefore, disaggregated into separate components for Australia, Canada, Mexico, Peru, Europe, the United States, and the rest of the world (RW).

In the second version of the model it is recognized that aggregate demand functions for zinc in each country, as specified in Model 1, may still fail to capture the differences in (1) technological aspects of the use of zinc in various industries, (2) the nature of growth phenomena in the different industries, and (3) the substitution possibilities between zinc and other materials, as dictated by the nature of the different industries.[2] In Model 2, therefore, these aspects are emphasized by subdivision of the total demand for zinc in each country according to end-uses of zinc—galvanizing, die castings, brass, rolled zinc, zinc oxide, and a miscellaneous category accounting for other minor uses of zinc. The rest of the system is the same as Model 1. Now, before turning to the specific details of the models and their estimation, some general remarks on the methodology used in preparing data and in estimation, which relate to both models, will be given in the next section.

General Remarks on Methodology and Data

The list of variables (table 5–5) at the end of this chapter describes in detail the definitions and sources of data of all the variables used in both versions of the model. However, certain aspects of some variables require more explanation.

Although the market-determined price of zinc in the world outside the United States is the LME price, the consumers' demand for zinc in a particular country will depend on the real cost of zinc in that country. Hence, the LME price used in the demand equations of the different countries was converted into local currencies and deflated by the local wholesale price indices. Similarly, the supply of zinc by producers of different countries will depend on the real revenues received from the sale of zinc in their local currencies. Thus, the price in the supply functions outside the United States is the LME price converted into local currencies and deflated by the local wholesale price indices. The demand and supply functions in the United States, however, are functions of the U.S. producers' price deflated by the U.S. wholesale price index.

Prices of many metals and materials (such as aluminum, copper, tin, and plastics) were tried to test the substitutability/complementarity of zinc. However, only the aluminum price showed some significant results. The LME price of aluminum was therefore selected, converted into local currencies, and deflated by the local wholesale price index, to represent the price of substitutes. Similarly, silver and lead appeared to be the chief coproducts, and hence their prices, converted to local currencies and deflated by the local wholesale price indices, were included in the supply equations.

Activity variables in the demand functions, as defined in table 5–5 are based on the discussion in the appendix to chapter 2.

Data on mine capacity were not available on a country basis. A five-year moving average of local mine production was therefore used as a proxy for the mine capacity variable.

Estimation Methods

As noted in the last chapter, the system is recursive (except for one equation for U.S. new scrap, which, however, plays a negligible part in the total system). Hence, the application of the ordinary least squares (OLS) method of estimation is justified. There is a problem associated with the presence of lagged dependent variables in the supply equations and the equation for the LME price, but it must be remembered that the presence of lagged dependent variables in our model is the result of a partial-adjustment hypothesis. In this case, then, application of the OLS will still yield consistent and efficient estimates of the parameters as long as the error terms are non-autocorrelated.[3] One, however, should not feel too confident about an absence of autocorrelation in such a large model based on time-series data. Usually one might also suspect the omission of some minor variables in the specification of the model, which, unless their effects totally cancel out in time-series data, would result in autocorrelated disturbances. Consequently, for all the equations of the model, the Cochrane-Orcutt method of correcting for autocorrelation has been applied along with the OLS estimation procedure.[4]

It is now necessary to turn to the models and the results of their estimation. It is to be noted that the sample data for estimation consist of annual observations for 1956–1974 for demand (except France), supply and trade equations, and 1960–1974 observations for demand in the case of France, and for stock and price equations. Additional data on lagged variables were supplied, where required. For all equations, R^2, DW (Durbin-Watson statistic), $\hat{\rho}$ (final estimate of the coefficient of autocorrelation in the Cochrane-Orcutt technique) and SER (standard error of estimate) are given. The t-test ratios for all coefficient estimates are given in parentheses below the corresponding estimate. *SUMLAG* represents the sum of the estimated coefficients of the lagged variables indicated in the parenthesis. The subscript t is omitted from the variables in the estimated equations for ease of presentation.

Model 1 Estimation Results

Model 1 is relatively aggregative on the demand side, as noted above. This model will now be presented, together with comments on each of the estimated equations. The discussion of the model will be divided into two parts, the U.S. subsystem and the rest of the world. As the free-market price enters as a determinant of the price of zinc in the U.S. subsystem, the results for the free-market system outside the United States will be presented first.

The FME World outside the United States

Although all the endogenous variables of this subsystem are determined within the subsystem itself, this subsystem is linked with the U.S. subsystem through interregional trade and exchange rates, and plays an important role in determining the long-run price behavior of the U.S. subsystem.

Demand (Consumption). Demand for zinc by consumers has been dealt with separately for Japan, the United Kingdom, West Germany, France, the rest of the developed world (R1), and the rest of the world (R2), for the reasons discussed above. All of the separate equations have polynomial-distributed lags (inverted V-shape), and the length and nature of the lags reflect the individual behavior patterns of the different countries. Many variations of the lag structures (including the exogenously determined weights on the inverted V-lag structure) were experimented with to capture the variations in the demand patterns of different countries. The most

Table 5–1
Estimates of Demand Equations: Model 1

Japan consumption

$$JCN = 1.68200 + 0.665452\,JA - 0.002711\,JLPZ_{-4}$$
(7.25) (24.54) (0.49)

$$- 0.005138\,JLPZ_{-5} - 0.007282\,JLPZ_{-6} - 0.009143\,JLPZ_{-7}$$
(0.70) (1.25) (1.65)

$$- 0.01072\,JLPZ_{-8}$$
(1.76)

$R^2 = 0.9946$ DW = 2.1501 $\hat{\rho} = -0.143526$ SER = 0.0485250
SUMLAG = −0.035(1.804)

U.K. consumption

$$KCN = 3.04151 + 0.419832\,KA - 0.0771857$$
(2.47) (3.01) (0.53)

$$(0.2\,KLPZ_{-1} + 0.4\,KLPZ_{-2} + 0.2\,KLPZ_{-3} + 0.2\,KLPZ_{-4})$$

$R^2 = 0.6512$ DW = 1.8233 $\hat{\rho} = 0.248701$ SER = 0.0396533

West Germany consumption

$$GCN = 1.41280 + 0.812471\,GA - 0.1084\,GLPZ_{-2} - 0.1746\,GLPZ_{-3}$$
(0.27) (6.75) (1.78) (1.92)

$$- 0.1986\,GLPZ_{-4} - 0.1803\,GLPZ_{-5} - 0.1198\,GLPZ_{-6}$$
(2.16) (2.45) (1.48)

$R^2 = 0.9014$ DW = 2.1622 $\hat{\rho} = -0.069566$ SER = 0.0764833
SUMLAG (GLPZ) = −0.78178 (2.38)

France consumption

$$FCN = -2.13178 + 0.544836\,FA - 0.0301\,FLPZ_{-3} - 0.0502\,FLPZ_{-4}$$
(1.21) (10.30) (0.96) (1.15)

$$-0.0604\,FLPZ_{-5} - 0.0504\,FLPZ_{-6} + 0.23562\,FLPL_{-3} + .32612\,FLPL_{-4}$$
(1.20) (0.92) (2.98) (3.36)

$$+ .41234\,FLPL_{-5} + .18562\,FLPL_{-6}$$
(2.87) (2.71)

$R^2 = 0.9146$ DW = 2.0481 $\hat{\rho} = -0.423832$ SER = 0.0374723
SUMLAG(FLPZ) = −0.25137 (1.21)
SUMLAG(FLPL) = 1.1597 (3.25)

Rest of the world (developed) consumption

$$R1CN = -2.64541 + 1.20933R1A - 0.00515(0.1R1LPZ_{-1} + 0.2R1LPZ_{-2}$$
(1.08) (15.15) (0.35)

$$+ \quad 0.3R1LPZ_{-3} + 0.25R1LPZ_{-4} + 0.1R1LPZ_{-5} - 0.05R1LPZ_{-6})$$

$$+ \quad 0.3512\,(0.1R1LPL_{-1} + 0.2R1LPL_{-2} + 0.3\,R1LPL_{-3} + 0.25R1LPL_{-4}$$
(2.11)

$$+ \quad 0.1R1LPL_{-5} + 0.05R1LPL_{-6})$$

$R^2 = 0.9828$ DW = 2.0070 $\hat{\rho} = -0.059317$ SER = 0.0418011

Rest of the world (less-developed countries) consumption

$$R2CN = \underset{(1.35)}{-3.87833} + \underset{(39.73)}{1.30784}R2A - \underset{(2.77)}{0.443118}(0.1R2LPZ_{-1} + 0.2R2LPZ_{-2} + 0.3R2LPZ_{-3} + 0.25R2LPZ_{-4} + 0.1R2LPZ_{-5} + 0.05R2LPZ_{-6}) + \underset{(1.90)}{0.969277}(0.1R2LPL_{-1} + 0.2R2LPL_{-2} + 0.3R2LPL_{-3} + 0.25R2LPL_{-4} + 0.1R2LPL_{-5} + 0.05R2LPL_{-6})$$

$R^2 = 0.9873$ DW $= 1.7276$ $\hat{\rho} = -0.25544$ SER $= 0.0588696$

---

successful lag structures were used in the estimations presented in table 5–1. In particular, Almon lag structures were more successful for Japan, West Germany, and France; whereas, some a priori specification of weights was required for the United Kingdom, R1, and R2. All equations are in double-logarithmic forms, so that the coefficients can be interpreted directly as elasticities.

As is clear from the equations in table 5–1, Japan has the slowest response in consumption to prices, the lagged effect starting only in the fourth year. Also, the elasticity of demand is very low (−0.035). In ascending order of magnitude of the elasticity of demand, the various countries can be ranked as R1 (−0.005), Japan (−0.035), the United Kingdom (−0.077), France (−0.25), R2 (−0.443), and West Germany (−0.78). Short-run elasticities, in the sense of current-year responses of consumption to changes in prices, are zero in all the countries, and, except for R1, R2 and the United Kingdom, the response of consumption to variations in the price of zinc does not start until two years after the date of consumption. This is quite expected because of the technological lag in adapting the plant for the use of new materials and for other reasons, as discussed previously. After the effect of variations in price have started, the mean lag for the effect varies from 2.6 years for Japan to 2.0 years for West Germany.

The importance of aluminum as the main substitute for zinc is revealed in the estimates. However, in the aggregative consumption equations, as aluminum is not substitutable in all uses of zinc, the cross-elasticity was not significant except in France (1.16), R1 (0.035), and R2 (0.97). A high cross-elasticity in France is expected as the French entrepreneurs (in the automobile sector mainly) were the first to start considering replacement of zinc by other metals. There could be two other reasons for the low cross-elasticities of demand: (1) as the history of the zinc industry reflects, the observed price of zinc may not have crossed the range of nonsubstitution; (2) part of the substitution effect is already absorbed in the coefficient of the variables representing the price of zinc itself because of its deflation by the wholesale price index. What the deflation does is (1) to relate the cost of zinc as an input to the price of the output in which it is used, and (2) to reflect its desirability relative to other inputs.

Primary Supply (Mine Production). For purposes of modeling the production of zinc, the world outside the United States is divided into Australia, Canada, Mexico, Peru, Europe, and the Rest-of-the World (RW), in order to reflect more effectively differences in cost structure. The response of producers to the price of zinc is assumed to be based on a partial adjustment hypothesis. It is further assumed (based on the estimation attempted and the technological nature of the lag structure) that producers' response to variations in prices in the same year is zero or negligible. Estimates of the supply equations presented in table 5–2 are based on double-logarithmic forms and hence the coefficients of the variables can be directly interpreted as elasticities.

Unlike demand, the mean lag on the supply side varies widely, ranging from only 0.35 for RW to 1.76 for Mexico. The mean lags for Australia, Canada, Peru, and Europe are 1.73, 1.05, 1.45, and 0.87, respectively. In these equations, however, it is somewhat difficult to conceptualize the price elasticities of supply as short-run and long-run elasticities. In the current year, as is clear from the above equations, producers' response to price is zero. Producers' response to price variations takes about a year in all cases. The supply response to P_{t-1} could be interpreted as a short-run elasticity. However, in the long run, as one would expect, capacity should generally be allowed to change—a change that would be induced by market conditions. But in the model, the capacity variable is treated as exogenous to the system. Also, capacities in the mineral industries take a long time to change, sometimes more than five or six years as has previously been discussed. This implies that the lagged adjustment of supply may be interpreted as depending on given levels of capacity. The elasticities, although not "true" long-run elasticities in the sense in which one is accustomed to think of them, do represent adjustments of the producers' supply to price variations over rather long periods of time. Further, it should be remembered that even if true long-run elasticities were available, which, in addition to investment (in plant) lags, should also reflect exploration lags, it is doubtful whether lags as long as ten to fifteen years would allow one to retain this model of pure competition based on perfect certainty. Henceforth, the term "elasticities" will be used to represent differences in the producers' response to prices, and "scale" effect to describe the adjustment of production to an exogenous change in the level of capacity.

The elasticities of supply (incorporating lagged adjustments) are fairly low, ranging from 0.24 for Australia to 0.617 for Peru. The elasticities for Canada, Mexico, Europe, and the rest of the world are 0.445, 0.284, 0.316, and 0.573, respectively. Higher elasticities for Peru, Canada, and the rest of

Table 5–2
Estimates of Supply Equations: Model 1

Australia production

$$AMP = -0.657108 + 0.152773\ AULPZ_{-1} + 0.366975\ AMP_{-1}$$
(1.50) (1.71) (1.52)

$$+\ 0.622726\ AMC$$
(2.82)

$R^2 = 0.9142$ $DW = 2.0853$ $\hat{\rho} = -0.284232$ $SER = 0.077347$

Canada production

$$CMP = 0.022502 + 0.227257\ CALPZ_{-1} = 0.488938\ CMP_{-1}$$
(0.41) (5.70) (6.82)

$$+ 0.63591\ CMC - 0.349647\ CAWG$$
(10.80) (2.14)

$R^2 = 0.9939$ $DW = 2.6626$ $\hat{\rho} = -0.41875$ $SER = 0.044985$

Mexico production

$$MMP = 0.044675 + 0.180808\ MELPZ_{-1} + 0.362909\ MMP_{-1}$$
(0.34) (2.51) (1.79)

$$+ 0.446713\ MMC$$
(2.42)

$R^2 = 0.6245$ $DW = 2.0441$ $\hat{\rho} = 0.308904$ $SER = 0.048678$

Peru production

$$PMP = -0.580186 + 0.365364\ PELPZ_{-1} + 0.407508\ PMP_{-1}$$
(0.82) (4.95) (2.84)

$$+\ 0.843738\ PMC - 0.491296\ PEWG$$
(5.77) (2.72)

$R^2 = 0.9774$ $DW = 2.1526$ $\hat{\rho} = -0.062617$ $SER = 0.0615007$

Europe production

$$EMP = 1.28495 + 0.146932\ EULPZ_{-1} + 0.178356\ EMC$$
(2.04) (2.16) (1.86)

$$+ 0.535682\ EMP_{-1} + 0.155198\ EUPSLD_{-1}$$
(3.60) (2.97)

$R^2 = 0.9532$ $DW = 1.885$ $\hat{\rho} = 0.194412$ $SER = 0.0298975$

Rest of the world production

$$RWMP = -0.250341 + 0.147002\ RWLPZ_{-1} + 0.743649\ RWMP_{-1}$$
(0.73) (2.09) (3.69)

$$+\ 0.168387\ RWMC$$
(1.1)

$R^2 = 0.9386$ $DW = 2.1234$ $\hat{\rho} = 0.14903$ $SER = 0.0542986$

the world, as compared to those for Europe, Mexico, and Australia, are consistent with the expectations, inasmuch as the mine deposits in the former set of countries are relatively newer than those in the latter set. The scale effect also corroborates this proposition, as it is 1.42, 1.24 and 0.66 for Peru, Canada, and the rest of the world, respectively, as compared with 0.98 for Australia, 0.7 for Mexico, and 0.38 for Europe.

Prices of coproducts were not found to be significant except for Europe, and hence they were excluded subsequently from the other supply equations.

Secondary Supply. Because of the limitations of data, only one equation (linear) was estimated for the supply of zinc from scrap (new plus old), for all countries together (the FME world excluding the United States).

$$\begin{aligned} RSCRAP = {} & \underset{(0.31)}{7.89815} + \underset{(0.22)}{0.00517}\, LMPZ_{-1} + \underset{(2.36)}{0.449838}\, RCNRES \\ & + \underset{(1.65)}{0.595480}\, TIME \end{aligned}$$

$$R^2 = 0.814 \qquad DW = 2.16 \qquad \hat{\rho} = -0.84069 \qquad SER = 0.068816 \tag{5.1}$$

As discussed earlier, the scrap has to compete with the primary resources available. An increase in available resources for a given level of consumption, as shown by the coefficient of RCNRES, reduces the need for recovery of metal from scrap. The variable *TIME* may be interpreted to include a compound effect of accumulation of scrap over time and a change in technology that may reduce the cost of recovery of metal from scrap. This is reflected in the positive and "significant" coefficient of *TIME*. As expected, the price variable does not seem to play any important role in the recovery of metal from scrap, as the coefficient is both small in magnitude and statistically insignificant. However, it is retained in the equation for the information that it provides.

Stocks. As argued earlier, stocks play an important role in the market for durable commodities. In the context of demand-supply forces, stocks for any time t represent excess supply in the market. Stocks may be derived from the identity

$$RSTK \equiv RSTK_{-1} + RMP + RSCRAP + RIMP - RCN - UIMP \tag{5.2}$$

However, stock identities, both for *RSTK* and *USTK* in this case, presented some unavoidable difficulties due to the inadequacy of data. The published figures for stocks did not match with the stock figures derived from identities.[5] Although either set cannot be claimed to be accurate, there are more doubts about the correctness of the stock figures derived from the identities. The figures for consumption over time have changed in their coverage with regard to (1) the inclusion of consumption from secondary resources in various countries (more distressingly, in different proportions), and (2) the inclusion of consumer stocks.

However, stocks are very important in the equations determining price, and certainly it would be impossible to treat stocks as exogenous to the system. Attempts were made to adjust the constituents of the identities based on the rough information available on the coverage of consumption in the various sources. The stock figures so derived were, however, not satisfactory. Alternatively, therefore, a behavioral equation for stocks was estimated which in turn was used in simulations in chapters 6 and 7. For inventory behavior in the free market, it was assumed that producers hold the stocks for day-to-day transactions purposes and speculative purposes (particularly stocks with the dealers at the LME and other places which are also included here). The stocks held for transaction purposes may be assumed to be a fixed proportion of their normal sales. However, the stocks held for speculative purposes may be assumed to be based on last year's prices. A rise in last year's prices may make producers/dealers expect a further rise in price in the current year, inducing them to hold larger stocks. An estimated linear equation based on such a hypothesis is

$$\begin{aligned} RSTK = & -265.068 + 2.3393\,LMPZ_{-1} + 0.958974\,RA \\ & \quad (3.33) \qquad (5.32) \qquad\qquad (1.73) \end{aligned}$$

$$R^2 = 0.8084 \quad DW = 1.5237 \quad \beta = 0.19778 \quad SER = 48.9835 \qquad (5.3)$$

Both coefficients are significant and support the hypothesis regarding producers' inventory behavior.

The LME Price (LMPZ). The determination of the free-market price was discussed in detail in the last chapter, and the discussion need not be repeated here. Besides the level and change in the U.S. government stockpile *GSA* in the last year, a dummy variable for the effective quota period (1961–1965) in the United States and the product of the dummy and the stockpile variable are also included in a linear price equation:

$$\begin{aligned} LMPZ = & \; 386.404 - 0.666768\,RSTKCN + 1.00063\,LMPZ_{-1} \\ & \quad (5.18) \qquad (2.21) \qquad\qquad (1.65) \\ & - 1375.84\,DUM - 2.56748\,\Delta GSA_{-1} \\ & \quad (2.07) \qquad\qquad (1.16) \end{aligned}$$

$$- 4.5005\ GSA_{-1} + 15.2408\ DUMGSA_{-1}$$
$$(2.27) \qquad (6.48)$$

$$R^2 = 0.9648 \quad DW = 2.5706 \quad \hat{\rho} = 0.141322 \quad SER = 15.52650 \qquad (5.4)$$

As expected, a rise in stocks relative to consumption depresses the current-year price. As was argued in chapter 4, a rise in the level of GSA_{-1} raises the potential for an increase in supply, and hence depresses the level of price as well. A rise in ΔGSA_{-1} adds to the effect of GSA_{-1}.

The U.S. Subsystem

The U.S. subsystem of demand and supply equations follows a format similar to the one discussed above in the case of the non-U.S. subsystem, and will therefore be discussed very briefly.

Demand (consumption). The equation for U.S. consumption is

$$UCN = 5.33155 + 0.830575\ UA - 0.1442\ USPZ_{-2} - 0.2293\ USPZ_{-3}$$
$$(2.74) \qquad (8.59) \qquad (2.00) \qquad (2.07)$$

$$- 0.2553\ USPZ_{-4} - 0.2221\ USPZ_{-5} - 0.1298\ USPZ_{-6}$$
$$(2.17) \qquad (2.21) \qquad (1.37)$$

$$R^2 = 0.9380 \quad DW = 2.3443 \quad \hat{\rho} = 0.46975 \quad SER = 0.475099$$
$$SUMLAG = -0.980674\ (0.4406) \qquad (5.5)$$

Compared to the demand equations in other countries, the long-run price elasticity of demand for zinc in the United States is quite high (fairly close to −1), though with a similar mean lag of about two years.

Primary Supply (mine production). The equation for U.S. production is

$$UMP = -1.23171 + 0.641603\ USPZ_{-1} + 0.248484\ UMP_{-1}$$
$$(1.54) \qquad (3.73) \qquad (1.73)$$

$$+ 0.577132\ UMC + 0.11234\ USPSLD_{-1} - 0.207118\ USWG$$
$$(3.82) \qquad (1.32) \qquad (1.30)$$

$$R^2 = 0.8103 \quad DW = 1.7656 \quad \hat{\rho} = -0.070708 \quad SER = 0.052855 \qquad (5.6)$$

The mean lag in the United States in the case of mine production is much longer (3.024) compared to the longest lag of 1.76 in other countries, indicating a slower adjustment process. However, the elasticity of supply is also the highest (0.854), compared to all the other areas of the world considered above. The scale effect is, on the average, quite similar to that in the rest of the world.

Secondary Supply. The supply of zinc from secondary sources in the United States has been divided into two categories: (1) supply from new scrap *UNS*, and (2) supply from old scrap *UOS*.

New Scrap. As discussed in the last chapter, zinc recovered from residues in the process of fabrication of metal products depends on the level of total metal fabrication and the price of zinc *USPZ*. Consumption of zinc in the United States *UCN* is used as a proxy for the former variable. The estimated linear relationship is

$$\underset{}{UNS} = \underset{(1.51)}{-13.375} + \underset{(1.69)}{0.16125}\, USPZ + \underset{(9.74)}{1.0515}\, UCN$$

$$R^2 = 0.9030 \quad DW = 0.88 \quad SER = 7.51866 \qquad (5.7)$$

Both variables *UCN* and *USPZ* are found to be important in explaining the supply recovered from new scrap, though *UCN* is more important.

Old Scrap. Recovery of zinc from old scrap in the United States, *UOS*, as argued for rest of the world, is hypothesized to depend on the ratio of consumption to resources *UCNRES*, the price of zinc *USPZ*, and a time trend *TIME*. The estimated linear relationship is

$$UOS = \underset{(1.29)}{-68.3927} + \underset{(1.93)}{0.13169}\, USPZ + \underset{(3.56)}{0.79545}\, UCNRES + \underset{(1.78)}{0.59548}\, TIME$$

$$R^2 = 0.814 \quad DW = 2.16 \quad \hat{\rho} = 0.084069 \quad SER = 0.068816 \qquad (5.8)$$

The corresponding elasticity estimates for *UCNRES* and *USPZ* are 0.11 and 0.65, respectively. The hypothesis seems to be reasonably satisfactory on the empirical grounds.

Stocks. The stock identity for the United States is as follows:

$$USTK = USTK_{-1} + UMP + UNS + UOS + UIMP - UCN - \Delta GSA \qquad (5.9)$$

As discussed in the case of *RSTK*, an equation for the inventory behavior of the U.S. producers was also estimated. In the case of the United States, where producers, along with government, try to regulate the market to keep prices stable through their stock holdings, the inventory behavior will be different from what it is in the free market. In this case, following the objective of price stabilization, U.S. producers may be expected to unload stocks on the market if the price rises or the activity level rises (which may put upward pressure on prices). U.S. government stocks play a complementary role in the quest for price stabilization. The estimated linear relationship supports these hypotheses:

$$\underset{}{USTK} = \underset{(10.00)}{1600.06} - \underset{(4.39)}{2.03927}\ USPZ - \underset{(8.75)}{6.14006}\ UA - \underset{(3.32)}{6.48776}\ GSA_{-1}$$

$$R^2 = 0.8648 \quad DW = 1.859 \quad \hat{\rho} = -0.499716 \quad SER = 24.8858 \qquad (5.10)$$

U.S. Producers' Price (USPZ). Since the process of determination of the U.S. price has been discussed earlier, only the results of the estimation will be noted here.

$$USPZ = \underset{(3.76)}{163.796} - \underset{(1.74)}{0.039345}\ USTKCN_{-1} + \underset{(4.88)}{0.40549}\ LMPZ$$

$$- \underset{(1.85)}{0.86804}\ UCAPUSE - \underset{(2.18)}{8.48345}\ DUM - \underset{(2.09)}{12.8180}\ DUMCLC$$

$$R^2 = 0.9353 \quad DW = 2.6376 \quad \hat{\rho} = -0.546844 \quad SER = 6.34357 \qquad (5.11)$$

The hypothesis about price determination in the administered market suggested $\Delta USTKCN$ and *UCAPUSE* as instruments used by the producers for price stabilization, and *LMPZ* as the long-run equilibrator. However, attempts to include $\Delta USTKCN$ did not succeed. Instead, $USTKCN_{-1}$ was found significant, and is included in equation 5.11. *DUM* and *DUMCLC* are, respectively, the dummy variables for the effective quota period and the effective period of wage and price freeze imposed by the Cost of Living Council in the United States. Rather than starting *DUMCLC* from August 1971 and ending in mid 1973, it was considered that the effect might be subject to a lag, rather than be immediate, and therefore *DUMCLC* began in 1972 and ended in 1973. Capacity utilization ratio turns out to be a better instrument compared to stocks for achieving the goal of price stabilization.

Interregional Trade. As discussed in the last chapter, it is proposed that the U.S. demand for imports (UIMP) depends on the price differential between the two markets and the variations in activity in the United States. *DUM* is the dummy variable for the effective quota period. Further, variations in the exchange rate between the United States and the United Kingdom also

Table 5–3
A Comparison of Price and Income Elasticities of Demand, Model 1 and Model 2

Country or Area	*Model 1*		*Model 2*	
	Price Elasticity	*Income Elasticity*	*Price Elasticity*	*Income Elasticity*
United States	−0.9807	0.831	−1.1688	0.8499
United Kingdom	−0.0772	0.4198	−0.2931	0.3763
Japan	−0.035	0.6651	−0.2334	0.7327
West Germany	−0.7818	0.8120	−1.2238	0.7805
France	−0.2514	0.5456	−0.6885	0.5454
R1	−0.0005	1.209	—	—
R2	−0.0443	1.308	—	—

influence variations in demand for imports. The estimated relation corroborates this hypothesis.

$$UIMP = \underset{(1.73)}{183.102} - \underset{(1.91)}{0.155449}\, LMUSPZ - \underset{(0.82)}{30.2352}\, ER - \underset{(1.52)}{30.0991}\, \mathrm{DUM} + \underset{(1.79)}{0.534956}\, UA$$

$$R^2 = 0.8038 \quad DW = 2.066 \quad \hat{\rho} = 0.130549 \quad SER = 65.27 \qquad (5.12)$$

Now attention will be turned to the estimation of Model 2.

Model 2 Estimation Results

As discussed earlier, the parameter estimates in the demand equations of Model 1 may not accurately reflect the technological and sectoral differences influencing the structure of demand for zinc. In fact, it has been theoretically established that aggregation in demand equations often results in downward bias in the estimates of price coefficients.[6] This is also corroborated by the estimates as given in table 5–3. Thus, in Model 2 an attempt has been made to eliminate the aggregation bias by disaggregating the total consumption of zinc by the major consumer countries into six sectors of demand.

Since the rest of Model 2 is the same as Model 1, only the estimated demand equations are presented for each end-use category (galvanizing, die castings, brass, rolled zinc, zinc oxide, miscellaneous) in table 5–4. In the simulations, where appropriate, Model 2 is used rather than Model 1.

Table 5–4
Estimates of Demand Equations: Model 2

United States

$$UCNG = \underset{(3.185)}{4.56751} + \underset{(8.882)}{0.61319}\,UAG - \underset{(1.500)}{0.08829}\,USPZ_{-2} - \underset{(1.591)}{0.14020}\,USPZ_{-3}$$
$$- \underset{(1.735)}{0.15570}\,USPZ_{-4} - \underset{(1.872)}{0.13490}\,USPZ_{-5} - \underset{(1.026)}{0.07774}\,USPZ_{-6}$$

$R^2 = 0.9200$ $DW = 2.1033$ $\hat{\rho} = 0.283707$ $SER = 0.04359$
SUMLAG $= -0.5969$ (0.322)

$$UCND = \underset{(2.277)}{9.29498} + \underset{(5.590)}{1.04590}\,UAD - \underset{(2.038)}{0.2857}\,USPZ_{-2} - \underset{(2.106)}{0.4603}\,USPZ_{-3}$$
$$- \underset{(2.194)}{0.5237}\,USPZ_{-4} - \underset{(2.232)}{0.4760}\,USPZ_{-5} - \underset{(1.654)}{0.3173}\,USPZ_{-6}$$

$R^2 = 0.8721$ $DW = 1.5614$ $\hat{\rho} = 0.66641$ $SER = 0.08816$
SUMLAG $= -2.06296$ (0.9209)

$$UCNB = \underset{(1.395)}{2.59349} + \underset{(7.947)}{0.71535}\,UAM - \underset{(0.934)}{0.09401}\,USPZ_{-2} - \underset{(0.913)}{0.1314}\,USPZ_{-3}$$
$$- \underset{(0.843)}{0.1120}\,USPZ_{-4} - \underset{(0.416)}{0.03604}\,USPZ_{-5} + \underset{(0.728)}{0.09661}\,USPZ_{-6}$$

$R^2 = 0.7745$ $DW = 2.3945$ $\hat{\rho} = -0.423346$ $SER = 0.110759$
SUMLAG $= -0.276824$ (0.417)

$$UCNR = \underset{(3.493)}{8.64763} + \underset{(0.961)}{0.115372}\,UAM - \underset{(1.53)}{0.1546}\,USPZ_{-3} - \underset{(1.61)}{0.2424}\,USPZ_{-4}$$
$$- \underset{(1.76)}{0.2623}\,USPZ_{-5} - \underset{(1.88)}{0.2172}\,USPZ_{-6} - \underset{(0.84)}{0.1043}\,USPZ_{-7}$$

$R^2 = 0.3599$ $DW = 2.0875$ $\hat{\rho} = 0.360169$ $SER = 0.07642$
SUMLAG $= -0.98169$ (0.525)

$$UCNO = \underset{(2.701)}{1.04988} + \underset{(9.625)}{0.786922}\,UACH$$

$R^2 = 0.8599$ $DW = 2.289$ $\hat{\rho} = 0.03979$ $SER = 0.084685$

$$UCNM = -\underset{(.544)}{21.0569} + \underset{(2.292)}{2.97622}\,UAM - \underset{(1.023)}{0.6171}\,USPZ_{-1} - \underset{(1.01)}{0.9465}\,USPZ_{-2}$$
$$-\underset{(0.976)}{0.9883}\,USPZ_{-3} - \underset{(0.855)}{0.7425}\,USPZ_{-4} - \underset{(0.305)}{0.2089}\,USPZ_{-5} + \underset{(3.112)}{1.829}\,USPL_{-1}$$
$$+ \underset{(2.842)}{2.578}\,USPL_{-2} + \underset{(2.06)}{2.247}\,USPL_{-3} + \underset{(0.61)}{0.8353}\,USPL_{-4} - \underset{(0.82)}{1.656}\,USPL_{-5}$$

$R^2 = 0.7713$ $DW = 1.209$ $\hat{\rho} = 0.805196$ $SER = 0.263022$
SUMLAG (USPZ) $= -3.50335$ (3.82)
SUMLAG (USPL) $= 5.8329$ (5.09)

$$UCN \equiv 0.3511\,UCNG + 0.3915\,UCND + 0.1074\,UCNB + 0.0353\,UCNR + 0.0898\,UCNO + 0.0249\,UCNM$$

Japan

$$JCNG = 1.35208 + 0.699227\,JAG$$
$$\qquad (9.85) \qquad (25.46)$$

$R^2 = 0.9862 \quad DW = 1.8949 \quad \hat{\rho} = 0.234793 \quad SER = 0.0675531$

$$JCND = 0.564936 + 0.906024\,JAD - 0.03445\,JLPZ_{-4} - 0.04708\,JLPZ_{-5}$$
$$(1.056) \qquad (15.057) \qquad (2.65) \qquad (2.73)$$
$$- 0.03788\,JLPZ_{-6} - 0.006853\,JLPZ_{-7} + 0.04599\,JLPZ_{-8}$$
$$(2.70) \qquad (0.52) \qquad (1.44)$$

$R^2 = 0.9896 \quad DW = 1.9394 \quad \hat{\rho} = 0.061477 \quad SER = 0.0973047$
SUMLAG = 0.0803 (0.047)

$$JCNB = 1.96904 + 0.641676\,JAM - 0.007136\,JLPZ_{-3} - 0.01187\,JLPZ_{-4}$$
$$(7.04) \qquad (17.746) \qquad (0.912) \qquad (1.16)$$
$$- 0.01419\,JLPZ_{-5} - 0.01411\,JLPZ_{-6} - 0.01163\,JLPZ_{-7}$$
$$(1.818) \qquad (2.195) \qquad (0.627)$$

$R^2 = 0.9773 \quad DW = 2.1094 \quad \hat{\rho} = -0.480644 \quad SER = 0.0803403$
SUMLAG = −0.05894 (0.0226)

$$JCNR = 3.44218 + 0.523270\,JAM - 0.004648\,JLPZ_{-4} - 0.01698\,JLPZ_{-5}$$
$$(7.162) \qquad (8.356) \qquad (0.412) \qquad (1.146)$$
$$- 0.03671\,JLPZ_{-6} - 0.06413\,JLPZ_{-7} - 0.09914\,JLPZ_{-8}$$
$$(3.267) \qquad (6.213) \qquad (3.498)$$

$R^2 = 0.9693 \quad DW = 1.826 \quad \hat{\rho} = -0.398641 \quad SER = 0.117456$
SUMLAG = −0.22152 (0.3508)

$$JCNO = 2.22392 + 0.477974\,JACH$$
$$(8.18) \qquad (8.84)$$

$R^2 = 0.9396 \quad DW = 1.8586 \quad \hat{\rho} = 0.41309 \quad SER = 0.0871077$

$$JCNM = 2.54542 + 1.52652\,JAM$$
$$(6.98) \qquad (20.64)$$

$R^2 = 0.9541 \quad DW = 2.069 \quad \hat{\rho} = -0.124968 \quad SER = 0.232825$

$$JCN \equiv 0.5651\,JCNG + 0.1783\,JCND + 0.1342\,JCNB + 0.0502\,JCNR + 0.0443\,JCNO + 0.0279\,JCNM$$

United Kingdom

$$KCNG = 7.50638 - 0.111559\,KAG - 0.7(0.1\,KLPZ_{-1} + 0.2\,KLPZ_{-2} + 0.3\,KLPZ_{-3}$$
$$(9.709) \qquad (0.66) \qquad \text{(a priori estimate)}$$
$$+ 0.25\,KLPZ_{-4} + 0.1\,KLPZ_{-5} + 0.05\,KLPZ_{-6})$$

$R^2 = 0.816 \quad DW = 1.51 \quad \hat{\rho} = 0.65331 \quad SER = 0.045185$

$$KCND = 1.35229 + 0.795387\,KAD - 0.140465(0.2\,KLPZ_{-1} + 0.4\,KLPZ_{-2}$$
$$(0.795) \qquad (6.481) \qquad (1.615)$$

Table 5–4 *(continued)*

$$+ 0.2\ KLPZ_{-3} + 0.2\ KLPZ_{-4}) + \underset{(.296)}{0.051255}\ (0.2\ KLPL_{-1} + 0.4\ KLPL_{-2}$$

$$+ 0.2\ KLPL_{-3} + 0.2\ KLPL_{-4})$$

$R^2 = 0.9558$ DW = 1.7545 $\hat{\rho} = 0.199387$ SER = 0.0351302

$$KCNB = \underset{(3.03)}{4.19539} + \underset{(0.78)}{0.289124}\ KAM - \underset{(2.384)}{0.1104}\ KLPM_{-1} - \underset{(2.01)}{0.1491}\ KLPM_{-2}$$

$$- \underset{(1.295)}{0.1160}\ KLPM_{-3} - \underset{(1.052)}{0.01108}\ KLPM_{-4} + \underset{(1.19)}{0.1656}\ KLPM_{-5}$$

$R^2 = 0.4877$ DW = 1.8017 $\hat{\rho} = 0.60985$ SER = 0.0732526
SUMLAG = −0.221 (0.404)

$$KCNR = \underset{(0.88)}{1.64541} + \underset{(2.57)}{0.358693}\ KAM - \underset{(1.67)}{0.04513}\ KLPZ_{-1} - \underset{(1.69)}{0.06520}\ KLPZ_{-2}$$

$$- \underset{(1.77)}{0.06021}\ KLPZ_{-3} - \underset{(1.27)}{0.03015}\ KLPZ_{-4} + \underset{(.5)}{0.02498}\ KLPZ_{-5} + \underset{(1.54)}{0.0802}\ KLPL_{-1}$$

$$+ \underset{(1.65)}{0.1237}\ KLPL_{-2} + \underset{(1.80)}{0.1304}\ KLPL_{-3} + \underset{(1.60)}{0.1005}\ KLPL_{-4} + \underset{(0.34)}{0.03378}\ KLPL_{-5}$$

$R^2 = 0.5226$ DW = 1.963 $\hat{\rho} = -0.573138$ SER = 0.0577758
SUMLAG (KLPZ) = −0.17572 (0.104)
SUMLAG (KLPL) = 0.46856 (0.262)

$$KCNO = \underset{(4.255)}{2.83169} + \underset{(3.00)}{0.406491}\ KACH$$

$R^2 = 0.8484$ DW = 2.045 $\hat{\rho} = 0.672055$ SER = 0.0666281

$$KCNM = \underset{(3.086)}{1.81977} + \underset{(4.77)}{0.595859}\ KAM$$

$R^2 = 0.7755$ DW = 2.002 $\hat{\rho} = 0.378766$ SER = 0.054522

$$KCN \equiv 0.2554\ KCNG + 0.1910\ KCND + 0.3451\ KCNB + 0.0658\ KCNR + 0.0712\ KCNO + 0.0715\ KCNM$$

West Germany

$$GCNG = \underset{(1.49)}{0.400624} + \underset{(5.496)}{0.938282}\ GAG$$

$R^2 = 0.9392$ DW = 1.4086 $\hat{\rho} = 0.679333$ SER = 0.0532663

$$GCND = \underset{(0.376)}{0.464648} + \underset{(12.15)}{1.35498}\ GAD - \underset{(1.393)}{0.06807}\ GLPM_{-4} - \underset{(1.525)}{0.1065}\ GLPM_{-5}$$

$$- \underset{(1.759)}{0.1153}\ GLPM_{-6} - \underset{(1.849)}{0.09448}\ GLPM_{-7} - \underset{(0.531)}{0.04403}\ GLPM_{-8}$$

$R^2 = 0.9717$ DW = 1.8083 $\hat{\rho} = 0.314765$ SER = 0.0833822
SUMLAG = −0.428395 (0.2282)

$$GCNB = \underset{(2.692)}{9.52026} + \underset{(4.121)}{0.839424}\ GAM - \underset{(1.923)}{0.2767}\ GLPZ_{-2} - \underset{(2.049)}{0.4393}\ GLPZ_{-3}$$

$$\underset{(2.254)}{-\,0.4878\,GLPZ_{-4}} \underset{(2.463)}{-\,0.4221\,GLPZ_{-5}} \underset{(1.300)}{-\,0.2423\,GLPZ_{-6}}$$

$R^2 = 0.7674$ DW = 1.6942 $\hat{\rho} = 0.257592$ SER = 0.174027
SUMLAG = 1.86833 (0.7697)

$$GCNR = \underset{(12.508)}{10.1385} + \underset{(1.300)}{0.08516\,GAM} \underset{(5.930)}{-\,0.2265\,GLPM_{-1}} \underset{(6.201)}{-\,0.3465\,GLPM_{-2}}$$

$$\underset{(6.464)}{-\,0.3598\,GLPM_{-3}} \underset{(5.429)}{-\,0.2666\,GLPM_{-4}} \underset{(0.9443)}{-\,0.06675\,GLPM_{-5}}$$

$R^2 = 0.7681$ DW = 2.006 $\hat{\rho} = -0.158593$ SER = 0.070506
SUMLAG = −1.2662 (0.2058)

$$GCNO = \underset{(3.787)}{8.77625} + \underset{(6.074)}{0.764031\,GACH} \underset{(2.092)}{-\,0.2172\,GLPZ_{-4}} \underset{(2.451)}{-\,0.3530\,GLPZ_{-5}}$$

$$\underset{(3.202)}{-\,0.4074\,GLPZ_{-6}} \underset{(3.829)}{-\,0.3804\,GLPZ_{-7}} \underset{(1.343)}{-\,0.2720\,GLPZ_{-8}}$$

$R^2 = 0.8909$ DW = 1.9826 $\hat{\rho} = -0.024556$ SER = 0.216633
SUMLAG = −1.62991 (0.415)

$$GCNM = \underset{(2.133)}{-110.993} + \underset{(2.255)}{3.55984\,GAM} \underset{(2.187)}{-\,1.241\,GLPZ_{-2}} \underset{(2.147)}{-\,1.918\,GLPZ_{-3}}$$

$$\underset{(2.050)}{-\,2.031\,GLPZ_{-4}} \underset{(1.754)}{-\,1.581\,GLPZ_{-5}} \underset{(0.724)}{-\,0.5660\,GLPZ_{-6}} + \underset{(2.451)}{3.786\,GLPL_{-2}}$$

$$+ \underset{(2.755)}{6.161\,GLPL_{-3}} + \underset{(3.119)}{7.126\,GLPL_{-4}} + \underset{(2.733)}{6.681\,GLPL_{-5}} + \underset{(1.215)}{4.8266\,GLPL_{-6}}$$

$R^2 = 0.8318$ DW = 2.3569 $\rho = 0.598144$ SER = 0.469015
SUMLAG (GLPZ) = −7.337 (3.873)
SUMLAG (GLPL) = 28.581 (9.376)

$$GCN \equiv 0.2552\,GCNG + 0.1150\,GCND + 0.3495\,GCNB + 0.2038\,GCNR + 0.0522\,GCNO + 0.0243\,GCNM$$

France

$$FCNG = \underset{(0.096)}{0.334972} + \underset{(1.291)}{0.921627\,FAG}$$

$R^2 = 0.5996$ DW = 1.9721 $\hat{\rho} = 0.730303$ SER = 0.0839508

$$FCND = \underset{(0.408)}{-2.34309} + \underset{(8.246)}{0.998936\,FAD} \underset{(0.386)}{-\,0.04047\,FLPZ_{-2}} \underset{(0.582)}{-\,0.09497\,FLPZ_{-3}}$$

$$\underset{(0.910)}{-\,0.1635\,FLPZ_{-4}} \underset{(1.44)}{-\,0.2460\,FLPZ_{-5}} \underset{(1.873)}{-\,0.3426\,FLPZ_{-6}}$$

$R^2 = 0.8888$ DW = 2.2725 $\hat{\rho} = -0.322375$ SER = 0.116632
SUMLAG (FLPZ) = −0.8876 (0.713)
SUMLAG (FLPL) = 1.381 (0.801)

$$FCNB = \underset{(0.564)}{1.27996} + \underset{(8.426)}{0.610529\,FAM} \underset{(2.371)}{-\,0.09777\,FLPZ_{-1}} \underset{(2.396)}{-\,0.1520\,FLPZ_{-2}}$$

$$\underset{(2.346)}{-\,0.1627\,FLPZ_{-3}} \underset{(1.920)}{-\,0.1299\,FLPZ_{-4}} \underset{(0.663)}{-\,0.05352\,FLPZ_{-5}} + \underset{(0.17)}{0.00785\,FLPL_{-1}}$$

Table 5–4 *(continued)*

$$+ 0.02319\,FLPL_{-2} + 0.09913\,FLPL_{-3} + 0.2180\,FLPL_{-4} + 0.3797\,FLPL_{-5}$$
$$(0.28) \quad (1.144) \quad (2.444) \quad (2.960)$$

$R^2 = 0.8611$ $DW = 2.4356$ $\rho = -0.649242$ $SER = 0.0636538$
SUMLAG (FLPZ) = −0.5959 (0.278)
SUMLAG (FLPL) = 0.7291 (0.353)

$$FCNR = 7.74115 + 0.137278\,FAM - 0.1038\,FLPZ_{-3} - 0.1689\,FLPZ_{-4}$$
$$(12.99) \quad (4.912) \quad (4.959) \quad (5.220)$$
$$- 0.1953\,FLPZ_{-5} - 0.1830\,FLPZ_{-6} - 0.1321\,FLPZ_{-7}$$
$$(5.493) \quad (5.316) \quad (3.337)$$

$R^2 = 0.6615$ $DW = 2.808$ $\hat{\rho} = -0.517516$ $SER = 0.0289703$
SUMLAG = −0.78317 (0.1424)

$$FCNO = 9.29635 + 0.347532\,FAM - 0.1538\,FLPZ_{-4} - 0.2592\,FLPZ_{-5}$$
$$(1.973) \quad (1.527) \quad (0.963) \quad (1.03)$$
$$- 0.3160\,FLPZ_{-6} - 0.3244\,FLPZ_{-7} - 0.2842\,FLPZ_{-8}$$
$$(1.127) \quad (1.264) \quad (1.271)$$

$R^2 = 0.2658$ $DW = 2.3849$ $\hat{\rho} = 0.525734$ $SER = 0.118496$
SUMLAG = −1.33765 (1.099)

$$FCNM = -0.678110 + 0.560293\,FAM - 0.1364\,FLPZ_{-1} - 0.2103\,FLPZ_{-2}$$
$$(0.152) \quad (3.714) \quad (1.711) \quad (1.703)$$
$$- 0.2218\,FLPZ_{-3} - 0.1709\,FLPZ_{-4} - 0.05670\,FLPZ_{-5} - 0.03173\,FLPL_{-1}$$
$$(1.622) \quad (1.264) \quad (0.363) \quad (0.296)$$
$$+ 0.02979\,FLPL_{-2} + 0.1846\,FLPL_{-3} + 0.4326\,FLPL_{-4} + 0.7739\,FLPL_{-5}$$
$$(0.185) \quad (1.076) \quad (2.467) \quad (3.154)$$

$R^2 = 0.6854$ $DW = 2.4373$ $\rho = -0.201393$ $SER = 0.0952469$
SUMLAG (FLPZ) = −0.797 (0.554)
SUMLAG (FLPL) = 1.3892 (0.698)

$$FCN = 0.2537\,FCNG + 0.1142\,FCND + 0.0159\,FCNB + 0.2881\,FCNR + 0.1670\,FCNO + 0.1613\,FCNM$$

Thus, in this chapter, the results of estimation of the two models are presented. The FME world market is divided into two subsystems: the U.S. market, which is administered by the producers and government in the United States; and the rest of the world, where the free-market system is predominant. Most of the results corroborate the hypotheses underlying the general formulations of the models in the last chapter. Whereas Model 1 entails aggregative data for demand equations, Model 2 disaggregates the demand side for five major consumers of zinc according to six end uses, and obtains a substantial improvement in the estimates of demand elasticities. Before application of these models to real world problems, the next chapter will test the validity of the model through the techniques of dynamic simulation.

Table 5–5
List of Variables

Name	*Description*	*Sources and Notes*
ALPZ*	LME prices of zinc in Australian dollars	Source: See AMP and ER
AMC	Australian mine capacity	A 5-year moving average of AMP
AMP*	Australia, mine production, recoverable zinc	Source: *Metal Statistics*, various issues
APWS	Australia, wholesale price index, domestic goods	UN monthly bulletin, various issues
AULPZ*	(ALPZ/APWS) * 100.0	
AUWG	(AWG/APWS) * 100.0	
AWG	Australia, wages in mining and quarrying	UN Yearbook of labor statistics, various issues
CAPLZ*	(CLPZ/CPWS) * 100.0	
CAWG	(CWG/CPWS) * 100.0	
CLPZ	LPZ in Canadian dollars	Source: See AMP
CMC	Canada, mine capacity	A 5-year moving average of CMP
CMP*	Canada, mine production	Source: Same as AMP
CPWS	Canada, wholesale price index, general	Source: See APWS
CWG	Canada, wages in metal mining	Source: See AWG
DUM	Dummy variable for the effective quota period in the United States	1 for 1961–1965, zero for other periods
DUMCLC	Dummy variable for the effective period of Cost of Living Council's wage and price freeze in the United States	1 for 1972–1973, zero for other periods
DUMGSA	(DUM * GSA)	
ELPZ*	Europe, price of zinc, same as GLPZ	See GLPZ
EMC	Europe, mine capacity	A 5-year moving average of EMP
EMP*	Europe, mine production, recoverable zinc	See AMP
EPSLD	Europe, a weighted average of silver and lead prices	Source: See ALPZ
EPWS	Europe, wholesale price index, weighted average of important producers in proportion of their production in 1963.	Source: See APWS
ER	(U.S. $/U.K. £)	Source: UN, International Financial Statistics
EULPZ*	(ELPZ/EPWS) * 100.0	
EUPSLD	(ELPSLD/EPWS) * 100.0	
EWG	Europe, wages in mining	Source: See AWG, a weighted index of wages in West Germany, Italy, Spain, and Sweden, weights 1963 zinc production.
FA	(0.3 * FAC + 0.3 * FAA + 0.4 * FAM)	
FAA	France, automobile production	Source: OECD, Main Economic Indicators, and UN, Statistical Yearbook, various issues

Note: an asterisk on a variable indicates that the variable is endogenous.

Table 5–5 *(continued)*

Name	*Description*	*Sources and Notes*
FAC	France, buildings construction	Source: OECD Main Economic Indicators, various issues, and UN Statistical Yearbook, various issues
FACH	France, Index of Chemical Manufactures Production	Source: Ibid.
FAD	(0.6 * FAA + 0.4 * FAM)	
FAG	(0.6 * FAC + 0.4 * FAM)	
FAM	France, production of manufactures, general	Source: See FAA
FCN*	(FCNG + FCND + FCNB + FCNR + FCNO + FCNM)	
FCNB*	France, consumption of zinc in brass	Source: International Lead and Zinc Study Group, (1960–1975)
FCND*	France, consumption of zinc in die-casting	Source: Ibid.
FCNG*	France, consumption of zinc in galvanizing	Source: Ibid.
FCNM*	France, consumption of zinc, miscellaneous	Source: Ibid.
FCNO*	France, consumption of zinc in zinc dust and oxides	Source: Ibid.
FCNR*	France, consumption of zinc in rolled zinc	Source: Ibid.
FLPL*	(FRLPL/FPWS) 100.0	Source: Ibid.
FLPZ*	(FRLPZ/FPWS) 100.0	Source: Ibid.
FMCAP	Free market world, zinc metal production capacity	Source: Ibid.
FMCN	Free market world, consumption of zinc	Source: See AMP
FMZP	Free market world, zinc metal production	Source: Ibid.
FPWS	France, wholesale price index, general	Source: See APWS
FRLPL	France, LME price of aluminum in francs	
FRLPM*	(FLPZ/FLPL) * 100.0	
FRLPZ*	France, LME price of zinc in francs	
GA	(0.3 * GAA + 0.3 * GAC + 0.4 * GAM)	
GAA	West Germany, automobile production	Source: See FAA
GAC	West Germany, buildings construction	Source: Ibid.
GACH	West Germany, production of chemical manufactures	Source: Ibid.
GAD	(0.6 * GAA + 0.4 * GAM)	
GAG	(0.6 * GAC + 0.4 * GAM)	
GAM	West Germany, production manufactures, general	Source: See FAA
GCN*	(GCNG + GCND + GCNB + GCNH + GCNO + GCNM)	
GCNB*	West Germany, consumption of zinc in brass	Source: See AMP
GCND*	West Germany, consumption of zinc in die casting	Source: Ibid.
GCNG*	West Germany, consumption of zinc in galvanizing	Source: Ibid.
GCNM*	West Germany, consumption of zinc, miscellaneous	Source: Ibid.
GCNO*	West Germany, consumption of zinc, oxides and dust	Source: Ibid.
GCNR*	West Germany, consumption of zinc in rolled zinc	Source: Ibid.
GLPL	(GRLPL/GPWS) * 100.0	Source: See ALPZ
GLPZ*	(GRLPZ/GPWS) * 100.0	Source: Ibid.
GPWS	West Germany, wholesale price index, general	Source: See APWS
GRLPL	Price of aluminum in D.M.	
GRLPM*	(GLPZ/GLPL) * 100.0	
GRLPZ*	LME price of zinc in D.M.	
GSA	United States, General Services Administration stockpile at the end of the years	Source: American Bureau of Metal

		Statistics, (1970–1976); McMahon et al. (1974, p. 74)
JA	(0.3 * JAA + 0.3 * JAC + 0.4 * JAM)	
JAA	Japan, automobile production	Source: See FAA
JAC	Japan, buildings construction	Source: Ibid. (estimated cost of construction in 1963 yen)
JACH	Japan, production of chemical manufactures	Source: Ibid.
JAD	(0.6 * JAA + 0.4 * JAM)	
JAG	(0.6 * JAC + 0.4 * JAM)	
JAM	Japan, production of manufactures, general	Source: See FAA
JCN*	(JCNG + JCND + JCNB + JCNR + JCNO + JCNM)	
JCNB*	Japan, consumption of zinc in brass	Source: See FCNB
JCND*	Japan, consumption of zinc in die castings	Source: Ibid.
JCNG*	Japan, consumption of zinc in galvanizing	Source: Ibid.
JCNM*	Japan, consumption of zinc, miscellaneous	Source: Ibid.
JCNO*	Japan, consumption of zinc, oxides and dust	Source: Ibid.
JCNR*	Japan, consumption of zinc in rolled zinc	Source: Ibid.
JLPL	(JPLPL/JPWS) * 100.0	Source: Ibid.
JLPZ*	(JPLPZ/JPWS) * 100.0	Source: Ibid.
JPLPL	LPL in Japanese currency	
JPLPM*	(JLPZ/JLPL) * 100.0	
JPLPZ*	LPZ in Japanese currency	
JPWS	Japan, wholesale price index, general	Source: See APWS
KA	(0.3 * KAA + 0.3 * KAC + 0.4 * KAM)	
KAA	United Kingdom, automobile production	Source: See FAA
KAC	United Kingdom, buildings construction	Source: Ibid.
KACH	United Kingdom, production of chemical manufactures	Source: Ibid.
KAD	(0.6 * KAA + 0.4 * KAM)	
KAG	(0.6 * KAC + 0.4 * KAM)	
KAM	United Kingdom, production of manufactures, general	Source: See FAA
KCN*	(KCNG + KCND + KCNB + KCNR + KCNO + KCNM)	
KCNB*	United Kingdom, consumption of zinc in die castings	Source: See FCNB
KCND*	United Kingdom, consumption of zinc in diecastings	Source: Ibid.
KCNG*	United Kingdom, consumption of zinc in galvanizing	Source: Ibid.
KCNM*	United Kingdom, consumption of zinc, miscellaneous	Source: Ibid.
KCNO*	United Kingdom, consumption of zinc, oxides and dust	Source: Ibid.
KCNR*	United Kingdom, consumption of zinc in rolled zinc	Source: Ibid.
KLPL	(UKLPL/KPWS) * 100.0	Source: Ibid.
KLPZ*	(UKLPZ/KPWS) * 100.0	Source: Ibid.
KPWS	United Kingdom, wholesale price index, general	Source: See APWS
LMPL	(LPL/KPWS) * 100.0	
LMPZ*	(LPZ/KPWS) * 100.0	
LPL	London Metal Exchange price of zinc in U.S. dollars	Source: See AMP
LPM	(LPZ/LPL) * 100.0	
LPZ*	London Metal Exchange price of zinc in U.S. dollars	Source: See AMP
LMUSPZ*	(LMPZ/USPZ) * 100.0	
MELPZ*	(MLPZ/MPWS) * 100.0	
MEWG	(MWG/MPWS) * 100.0	
MLPZ*	LPZ in Mexican currency	

Table 5–5 *(continued)*

Name	*Description*	*Sources and Notes*
MMC	Mexico, mine capacity	Source: 5-year moving average of MMP
MMP*	Mexico, mine production, recoverable zinc	Source: See AMP
MPWS	Mexico, wholesale price index, general (Mexico city)	Source: See APWS
MWG	Mexico, wages in mining and quarrying	Source: See AWG
PELPZ*	(PLPZ/PPWS) * 100.0	
PEWG	(PWG/PPWS) * 100.0	
PLPZ*	LPZ in Peruvian Currency	
PMC	Peru, mine capacity	Source: 5-year moving average of PMP
PMP*	Peru, mine production	Source: See AMP
PPWS	Peru, wholesale price index, weighted average of building materials and farm products, equal weights	Source: See APWS
PWG	Peru, wages in mining	Source: See AWG (Consumer price index used as a proxy for PWG)
RA	Rest-of-the-world (Freeworld less United States) industrial production	Source: UN Statistical Yearbook, various issues
R1A	Rest-of-the-world, industrial production	Source: Ibid. (weighted average of EFTA, Oceania, Canada indices of industrial production, weights in proportion of their consumption of zinc in 1963)
R2A	Rest-of-the-world, industrial production, developing countries	Source: Ibid.
RCN*	(JCN + KCN + GCN + FCN + R1CN + R2CN)	
RCNRES*	(RCN/WRES)	
R1CN*	Rest of the developed free world [Canada + (Europe − UK − France − West Germany) + Australia + S. Africa] consumption of zinc	Source: See AMP
R2CN*	Rest-of-the-world (developing countries) FMCN − UCN − KCN − JCN − GCN − FCN − R1CN	Source: Ibid.
R1LPL	(LPL/R1PWS) * 100.0	
R2LPL	(LPL/R2PWS) * 100.0	
R1LPM*	(R1LPZ/R1LPL) * 100.0	
R2LPM*	(R2LPZ/R2LPL) * 100.0	
R1LPZ*	(LPZ/R1PWS) * 100.0	
R2LPZ*	(LPZ/R2PWS) * 100.0	
RMC	Rest-of-the-world (free world less United States) mine capacity	Source: 5-year moving average of RMP
RMP*	(AMP + CMP + MMP + PMP + EMP + RWMP)	
ROWG	(RWWG/RWPWS) * 100.0	
R1PWS	Rest of the developed world, wholesale price index, same as EPWS	
R2PWS	Rest-of-the-world (free world—developing countries) wholesale price index, same as UPWS	Because of too many problems of aggregation and data availability, US price index was used as a proxy variable.

RSCRAP*	Rest-of-the-world (free world less United States), old and new scrap	Source: See AMP
RSTK*	Rest-of-the-world (free world less United Stats), producers' stocks, end of year	Source: International Lead & Zinc Study Group *Bulletin*, various issues
RSTKCN*	(RSTK/RCN)	
RWLPZ*	(LPZ/RWPWS) * 100.0	
RWMC	Restof-the-world (See RWMP)	5-year moving average of RWMP
RWMP*	Rest-of-the-world (free world less United States, Australia, Canada, Europe, Mexico, and Peru) mine production	Source: See AMP
RWPWS	Rest-of-the-world (see RWWG), wholesale price index	Source: See APWS, a weighted average of Japan and Zaire prices (see RWWG)
RWWG	Rest-of-the-world (free world minus United States, Canada, Australia, Mexico, Peru, Europe)	Source: See AWG A weighted index of wages in mining in Japan and Zaire. Weights in proportion of production of zinc in 1963.
RZP	(FMZP − USZP) price index	Source: See GSA
TIME	1963 = 100, increasing (decreasing) by 2 each succeeding (preceding) year.	
UA	(0.3 * UAA + 0.3 * UAC + 0.4 * UAM)	
UAA	United States, automobile production	Source: See FAA
UAC	United States, buildings construction	Source: Ibid.
UACH	United States, produciton of chemical manufactures	Source: Ibid.
UAD	(0.6 * UAA + 0.4 * UAM)	
UAG	(0.6 * UAC + 0.4 * UAM)	
UAM	United States production of durable manufactures, general	
UCAPUSE	(USZP/USCAP)	Source: See GSA
UCN*	(UCNG + UCND + UCNB + UCNR + UCNO + UCNM)	
UCNB*	United States, consumption of zinc in brass	Source: see AMP
UCND*	United States, consumption of zinc in die castings	Source: Ibid.
UCNG*	United States, consumption of zinc in galvanizing	Source; Ibid.
UCNM*	United States, consumption of zinc, miscellaneous	Source: Ibid.
UCNO*	United States, consumption of zinc, oxides including the oxides directly manufactured from ores	Source: Ibid.
UCNR*	United States, consumption of zinc in rolled zinc	Source: Ibid.
UCNRES*	(UCN/WRES)	
UIMP*	United States, net imports	Source: See AMP
UKLPL	LPL in U.K. pounds	
UKLPM*	(KLPZ/KLPL) * 100.0	
UKLPZ*	LPZ in U.K. pounds	
ULMPZ*	(ULPZ/UPWS) * 100.0	
ULPL	LME price of aluminum	Source: See AMP
ULPM*	(ULPZ/ULPL) * 100.0	
ULPZ*	LME price of zinc	Source: See AMP
UMC	United States, mine capacity	5-year moving average of UMP

Table 5–5 *(continued)*

Name	*Description*	*Sources and Notes*
UMP*	United States, mine production, recoverable zinc	Source: See AMP
UNS*	United States, production of zinc from new scrap	Source: See GSA
UOS*	United States, production of zinc from old scrap	Source: See GSA
UPL	United States price of aluminum	Source: See AMP
UPSLD	Weighted average of silver and lead prices in the United States	Source: See GSA Weights: silver = 0.3 lead = 0.7
UPWS	United States, wholesale price index, general	Source: See APWS
UPZ*	United States prime western price of zinc	Source: See AMP
USCAP	United States, zinc metal production capacity	Source: See GSA
USLMPL	LME price of aluminum in U.S. dollars	
USPL	(UPL/UPWS) * 100.0	
USPM*	(UPZ/UPL) * 100.0	
USPSLD	(UPSLD/UPWS) * 100.0	
USPZ*	(UPZ/UPWS) * 100.0	
USTK*	United States, producers stock, end of year	Source: See RSTK
USTKCN*	(USTK/UCN)	
USWG	(UWG/UPWS) * 100.0	
USZP	United States, zinc metal production	Source: See GSA
UWG	United States, wages in metal mining	Source: See AWG
WRES	Free-market world economic resources containing zinc	Source: Cammorata (1975, p. 6)

Notes

1. For the sake of convenience, the first version and the second version of the model have been referred to in the text as Model 1 and Model 2, respectively.
2. For a detailed discussion of these aspects, see appendix to chapter 2.
3. For a discussion and proof of this proposition, see Johnston (1972, pp. 300–320).
4. For details, see Cochrane and Orcutt (1949), and Johnston (1972, pp. 261–262).
5. This is, however, not a problem peculiar to the zinc industry alone. The problem has been faced by many other researchers; for example, Fisher, Cootner, and Baily (1972), and Labys (1973).
6. For a brief but comprehensive discussion of aggregation bias and associated prediction errors, see Gupta (1969, pp. 1–6). Gains from disaggregation are well discussed in Orcutt, Watts, and Edwards (1968).

6 Performance of the Models of the World Zinc Industry

Methodological Remarks

Specification and estimation of the econometric models of the world zinc industry have been based on the hypothesis of two market systems. The U.S. producers of zinc are assumed to follow the goal of price stabilization, supported by the U.S. government. The decision making process in the rest of the world market for zinc is assumed to operate in a competitive framework, with the London Metal Exchange as its base. Although the individual equations seem to perform well, the validity of the model as a whole depends on the simultaneous interactions of the various equations which constitute the complete model.

A hypothesis concerning a system can be tested by assessing its ability to predict the actual behavior of the system. Usually this may be done by considering the degree to which the predicted observations from the model depart from the actual observations.[1] Predictions could either be ex post (that is, retrospective predictions over the sample period) or ex ante (prospective predictions beyond the sample period). Ex ante predictions require forecasting of all the relevant exogenous variables, and this itself may produce errors. These errors may become compounded in the ex ante predictions of the endogenous variables, and may therefore not yield an accurate test of the model. Thus, ex post predictions, also called base simulations, have been selected.

Base simulations may be one-period similations or dynamic simulations over several periods. The one-period simulations use the actual values of the lagged endogenous variables. For example, consider a system such as:

$$AX_t + BY_t + \sum_{j=1}^{T} B_j Y_{t-j} + CZ_t + D = U_t, \qquad (6.1)$$

where X_t is an $m \times 1$ vector of exogenous variables; Y_t is an $n \times 1$ vector of endogenous variables; Z_t is a $q \times 1$ vector of policy instruments; U_t is an $n \times 1$ vector of stochastic disturbance terms; and A, B, C, and D are coefficient matrixes whose parameters have been estimated by standard econometric techniques. Rearranging the matrices and vectors yields the base solution (assuming t varies over the sample period used in estimating the system) for all the endogenous variables, that is

$$Y_t = -B^{-1}AX_t - B^{-1} \sum_{j=1}^{T} B_j Y_{t-j} - B^{-1}CZ_t - B^{-1}D + B^{-1}U_t \qquad (6.2)$$

If the Y_{t-j} are the actual values of the sample period then the simulation is called a one-period base simulation. The simulations are dynamic if the lagged endogenous values Y_{t-j} for any simulation period t are predicted values of Y (though in the beginning year of simulation, actual lagged values are used). Obviously, dynamic simulations provide a more rigorous test of the model than simulations in which the values of the lagged endogenous variables are not generated from the structure of the model itself. The evaluation of these models is based on the rigorous test of dynamic simulations.[2]

Summary results of these dynamic simulations are given in table 6–1. Two summary measures have been selected by averaging over the ten years the percentage deviation (P.DEV) values from tables 6–2 and 6–3 (at the end of this chapter). The first is the average percentage absolute deviation, which is the average of the absolute deviations of solution values from actual values, expressed as percentages of actual values (APAD). This measures failure of the model to reproduce the historical data in percentage terms. However, this measure does not indicate very much about the tendency of the model, in general, to overestimate or underestimate the actual values. For example, there may be a situation where the short-term fluctuations are overestimated or underestimated but, on the average for the whole simulation period, the predicted values are very near to the long-term trend in the actual data. This type of prediction error may be summarized by the algebraic average of the percentage (that is, in averaging, signs are taken into consideration). A larger negative (positive) value of the average percentage deviation (APD) will reflect the downward (upward) bias in the predicted values over the simulation period. A value of APD closer to zero may be used to reflect an absence of long-run bias in the predicted values. The summary measures of both types for all the endogenous variables of both models are contained in table 6–1. Since results of the simulations for the two models are, in general, quite similar, only the results of the simulations for Model 1 are discussed in more detail. The results of the simulations, in detail, for Models 1 and 2 are presented in tables 6–2 and 6–3, respectively. In the next two sections, these simulation results for both the models will be discussed.

Model 1 Simulations

The performance of both models is reasonably satisfactory. While consumption, production, and trade do not have large prediction errors, prediction errors for stock variables do not seem to be very satisfactory. The error in the prediction of stocks is to a great degree transmitted to error in the

Table 6–1
Summary Measures of Sample Period Dynamic Simulation Errors (1965–1974): Models 1 and 2

	Model 1		*Model 2*	
Variables	*Percentage Absolute Deviation (APAD)*[a]	*Percentage Algebraic Deviation (APD)*[b]	*Percentage Absolute Deviation (APAD)*[a]	*Percentage Algebraic Deviation (APD)*[b]
Consumption				
JCN	3.70	0.57	4.39	0.15
KCN	3.03	0.48	2.65	0.59
GCN	4.53	0.02	3.96	2.57
FCN	2.64	−0.23	4.33	0.43
R1CN	3.37	0.68	3.37	0.68
R2CN	3.26	0.02	3.26	0.02
RCN	4.48	4.15	4.72	4.56
UCN	3.83	−0.69	3.53	−1.21
Production				
AMP	6.43	4.84	6.43	4.84
CMP	10.91	−10.91	10.91	−10.91
MMP	4.44	−0.91	4.44	0.91
PMP	6.03	−6.03	6.03	−6.03
EMP	2.38	−0.70	2.38	−0.70
RWMP	4.33	−3.01	4.33	−3.01
RMP	4.64	−4.44	4.64	−4.44
RSCRAP	7.79	−7.79	7.79	−7.79
UMP	3.20	−0.52	3.20	−0.54
UNS	5.07	−2.33	5.07	−2.33
UOS	10.50	−10.50	10.50	−10.50
Stocks				
RSTK	39.7	13.22	39.50	13.40
USTK	17.63	−2.74	17.63	−2.74
Prices				
LMPZ	12.39	−0.24	12.89	−0.01
USPZ	5.62	−0.95	5.54	−0.86
Trade				
UIMP	5.18	−0.596	5.19	−0.61

[a]Calculated from $\frac{1}{10} \sum_{i=1}^{10} \frac{|\text{Predicted} - \text{Actual}|}{\text{Actual}} * 100.0$ (without algebraic sign)

[b]Calculated from $\frac{1}{10} \sum_{i=1}^{10} \frac{(\text{Predicted} - \text{Actual})}{\text{Actual}} * 100.0$ (with algebraic sign)

prediction of prices, though the latter is still within reasonable bounds. In this section, a focus will be placed on the simulation results of Model 1.

Consumption

Dynamic simulation results for consumption are very good, and the best among all the variables of the model. The average percentage absolute error

(APAD) ranges between 2.64 for France and 4.53 for West Germany. The higher figure for West Germany is due to the failure of the model to capture a sharp drop in consumption in 1969. However, none of the consumption variables drift outside the range of the actual value ± 0.7 percent. In general, except for France and the United States, the drifts have been on the positive side, which is reflected in the compounded drift of 4.15 in the total non-U.S. consumption *RCN*.

Production

Dynamic simulations for primary and secondary supplies of zinc are reasonably satisfactory. The APAD for primary production ranges between 2.38 for Europe to 10.91 for Canada. The large errors in the simulated values for Canadian mine production are attributable to a number of discoveries of new zinc mines in the province of Ontario during the simulation period. This is revealed in the negative value of the APD, which implies an underestimation of the actual values throughout the simulation period. The same is true in the case of mine production in Peru. Australia, on the other hand, had a number of mine closures (obsolete mines) during the simulation period, which resulted in an overestimate of the actual mine production. Both the APAD and APD in the case of mine production for Mexico, Europe, and the United States are quite satisfactory.

The secondary supply, in general, has been underestimated by the model. While metal recovery from new scrap in the United States shows an APAD within reasonable limits (about 5 percent), the predicted values have slightly downward bias. The old scrap in the United States, on the other hand, has been poorly predicted. More disturbing is the fact that the errors in the long run do not average out, but show a downward drift by 10.5 percent. The only obvious reason for this drift seems to be the inability of the model to capture the effect of the cumulative pile of scrap on the ground, over time. Also, this is certainly more important for the simulation period as compared to the other parts of the sample period, as it has a cumulative effect. The same seems to be true, although to a lesser degree, for the predicted values of the metal recovery from scrap in the rest of the world.

Stocks

Errors in the prediction of stocks usually reflect the accumulated errors in the prediction of consumption and production. Besides, stocks as a variable is very sensitive to adjustments and expectations. As was argued earlier, stocks in the model used for simulation appear in the form of inventory behavior of the producers/dealers of zinc rather than the residuals between consumption and production. The rest-of-the-world stocks (*RSTK*) have very large prediction errors, up to 40 percent. These large errors, however, may be attributed to (1) inadequacy of the published data for the non-U.S. world, and (2) the actual inventory behavior of producers, not well captured by the model.

On examining the prediction errors more closely, one finds that three predicted observations (1965, 1966, and 1973) account for more than 70 percent of the total error (APAD) reported in the stock variable. During these years, as noted earlier, the world zinc industry observed a high instability of price at the LME. For example, the daily quotation of prices during 1964 alone rose by more than 100 percent. In such abnormal circumstances, producers may have deviated from the normal expectations mechanism specified in the model. The price increase in 1964, rather than creating expectations of a further rise in price, seems to have induced the producers to expect a decrease in price in the future. This probably encouraged speculators to unload their stocks onto the market, which may have resulted in a large overestimate of stocks in 1965. The same is true for the very abnormal year of 1973, when the daily prices at the LME rose by more than 200 percent. If we take out these two abnormal years, where some perverse expectations may have played a key role, the error in stock figures (APAD) is reduced to less than half of the reported figure for the total simulation period.

The data accuracy for stocks is much better for the United States. The error in prediction is also less than half the size observed in the rest of the world. It may be recalled that the stocks in the United States are also used as an instrument by the producers for stabilization of U.S. prices. However, the abnormal year 1965 coincided with about one-third of the total APAD for the simulation period. Also in that year, for the first time, the U.S. government released stockpiles of about 70,000 tons, which seem to have worked as complementary to the efforts of the producers in their goal of price stabilization. This may have reduced the need for decreasing the producers' stocks, resulting in a large underestimate of stocks by the model during this year.

Prices

The LME price contains large prediction errors as compared to the U.S. price for at least two reasons: (1) errors in the stocks in the free market are much more than in the U.S. market, and (2) the free-market price is much more unstable than the U.S. price. However, the comforting element in the price behavior in both markets is that the predictions are quite satisfactory on the APD criterion. The drift of prices from the long run trend is less than 0.6 percent over time. This implies that the model is more successful in predicting relatively long-term price behavior as compared to short-term movements.

Trade

The errors in prediction of interregional trade are reasonably satisfactory. The APD is also less than 1 percent.

Model 2 Simulations

As noted above, the only difference in Model 2 from Model 1 is that Model 2 explicitly incorporates the differences in consumption structure within the various countries. These structural differences, again, may be attributed to the differences in the stage of development, pattern of industrial production, and technological differences and/or preferences in the use of zinc in the various countries. In the estimates of Model 2 it was observed that Model 2 made definite improvements in the estimates of the parameters of the consumption demand equations in almost all the countries. However, in simulation, predictions from Model 2 are not substantially different from those of Model 1 (see table 6–1). There are larger absolute errors in Japan and France than were observed in Model 1. For West Germany, the United States, and the United Kingdom, the model shows some improvements in the predicted values, though long-term bias is somewhat larger. Since the difference is so small, it is hard to think of any valid reasons for the differences in the prediction of the two models. These small differences in prediction errors in the two models are also translated into small differences in the prediction errors in the prices. In general, on the grounds of these empirical tests of validity, it is difficult to select one model as better than the other. However, both models are needed because some policy changes can more aptly be incorporated into one model than the other. Since both models are similar on tests of validity, one is left free to use the one which is more suitable for any given policy change.

Table 6–2
Results of Dynamic Simulation: Model 1

THE WORLD ZINC INDUSTRY
DYNAMIC SIMULATION 1965-1974
MODEL ONE

VAR. NO. 1 IS RCN

	ACTUAL	SOLN	DEV	P.DEV
1965	109.545	117.607	8.062	7.36
1966	112.811	122.087	9.276	8.22
1967	119.096	124.701	5.605	4.71
1968	131.247	134.741	3.494	2.66
1969	146.669	144.266	-2.403	-1.64
1970	148.066	151.103	3.037	2.05
1971	147.240	163.396	16.156	10.97
1972	164.650	172.970	8.320	5.05
1973	181.993	185.461	3.468	1.91
1974	172.762	173.093	.331	.19

VAR. NO. 2 IS UCN

	ACTUAL	SOLN	DEV	P.DEV
1965	122.099	117.650	-4.449	-3.64
1966	127.256	117.105	-10.151	-7.98
1967	111.750	115.366	3.616	3.24
1968	120.681	122.726	2.045	1.69
1969	119.567	122.162	2.595	2.17
1970	108.315	113.679	5.364	4.95
1971	113.453	121.694	8.241	7.26
1972	127.505	126.146	-1.359	-1.07
1973	134.963	130.737	-4.226	-3.13
1974	116.713	120.614	3.901	3.34

VAR. NO. 3 IS JCN

	ACTUAL	SOLN	DEV	P.DEV
1965	108.139	116.844	8.705	8.05
1966	127.568	126.174	-1.394	-1.09
1967	151.559	148.397	-3.162	-2.09
1968	171.546	172.264	.718	.42
1969	196.882	191.021	-5.861	-2.98
1970	204.496	209.176	4.680	2.29
1971	204.824	222.397	17.573	8.58
1972	235.215	232.576	-2.639	-1.12
1973	267.443	245.054	-22.389	-8.37
1974	222.678	227.155	4.477	2.01

VAR. NO. 4 IS KCN

	ACTUAL	SOLN	DEV	P.DEV
1965	102.082	104.576	2.494	2.44
1966	100.281	102.794	2.513	2.51
1967	95.527	102.430	6.903	7.23
1968	103.910	104.883	.973	.94
1969	107.539	105.124	-2.415	-2.25
1970	104.163	103.583	-.580	-.56
1971	100.731	103.170	2.439	2.42
1972	102.785	104.863	2.078	2.02
1973	111.111	103.783	-7.328	-6.60
1974	99.297	95.962	-3.335	-3.36

Table 6–2 *(continued)*

THE WORLD ZINC INDUSTRY
DYNAMIC SIMULATION 1965-1974
MODEL ONE

(CONTINUED)

VAR. NO. 5 IS GCN

	ACTUAL	SOLN	DEV	P.DEV
1965	116.967	116.761	-.206	-.18
1966	111.557	115.820	4.263	3.82
1967	104.809	102.492	-2.317	-2.21
1968	117.404	107.242	-10.162	-8.66
1969	101.940	114.179	12.239	12.01
1970	123.415	119.432	-3.983	-3.23
1971	124.290	127.565	3.275	2.64
1972	127.295	132.748	5.453	4.28
1973	137.787	134.519	-3.268	-2.37
1974	121.503	114.360	-7.143	-5.88

VAR. NO. 6 IS FCN

	ACTUAL	SOLN	DEV	P.DEV
1965	105.876	103.514	-2.362	-2.23
1966	105.195	106.277	1.082	1.03
1967	108.277	105.274	-3.003	-2.77
1968	100.502	105.669	5.167	5.14
1969	112.218	107.276	-4.942	-4.40
1970	108.026	107.927	-.099	-.09
1971	110.247	114.705	4.458	4.04
1972	124.686	126.329	1.643	1.32
1973	138.302	139.001	.699	.51
1974	133.286	126.778	-6.508	-4.88

VAR. NO. 7 IS R1CN

	ACTUAL	SOLN	DEV	P.DEV
1965	108.888	109.907	1.019	.94
1966	110.323	116.009	5.686	5.15
1967	116.687	118.337	1.650	1.41
1968	123.065	126.119	3.054	2.48
1969	139.087	134.584	-4.503	-3.24
1970	147.990	138.225	-9.765	-6.60
1971	133.903	145.602	11.699	8.74
1972	153.673	155.971	2.298	1.50
1973	168.833	168.285	-.548	-.32
1974	171.962	166.300	-5.662	-3.29

VAR. NO. 8 IS R2CN

	ACTUAL	SOLN	DEV	P.DEV
1965	110.614	112.672	2.058	1.86
1966	116.170	117.482	1.312	1.13
1967	125.700	124.284	-1.416	-1.13
1968	147.516	138.143	-9.373	-6.35
1969	156.594	153.013	-3.581	-2.29
1970	178.726	167.962	-10.764	-6.02
1971	188.347	197.370	9.023	4.79
1972	209.982	212.098	2.116	1.01
1973	231.030	247.433	16.403	7.10
1974	235.501	233.374	-2.127	-.90

THE WORLD ZINC INDUSTRY
DYNAMIC SIMULATION 1965-1974
MODEL ONE

(CONTINUED)

VAR. NO. 44 IS RMP

	ACTUAL	SOLN	DEV	P.DEV
1965	118.979	120.195	1.216	1.02
1966	128.582	126.660	-1.922	-1.49
1967	139.412	131.154	-8.258	-5.92
1968	145.556	137.944	-7.612	-5.23
1969	156.276	145.552	-10.724	-6.86
1970	160.459	149.645	-10.814	-6.74
1971	160.876	154.365	-6.511	-4.05
1972	167.687	159.767	-7.920	-4.72
1973	173.418	166.283	-7.135	-4.11
1974	169.831	159.136	-10.695	-6.30

VAR. NO. 45 IS UMP

	ACTUAL	SOLN	DEV	P.DEV
1965	115.476	112.613	-2.863	-2.48
1966	108.186	110.939	2.753	2.54
1967	103.812	106.329	2.517	2.42
1968	100.042	99.928	-.114	-.11
1969	104.520	98.685	-5.835	-5.58
1970	100.937	96.737	-4.200	-4.16
1971	94.959	96.100	1.141	1.20
1972	90.377	96.845	6.468	7.16
1973	90.481	86.520	-3.961	-4.38
1974	93.439	91.596	-1.843	-1.97

VAR. NO. 46 IS AMP

	ACTUAL	SOLN	DEV	P.DEV
1965	98.361	109.324	10.963	11.15
1966	104.918	113.510	8.592	8.19
1967	113.115	117.402	4.287	3.79
1968	118.033	125.217	7.184	6.09
1969	142.623	132.259	-10.364	-7.27
1970	136.066	139.906	3.840	2.82
1971	127.869	148.934	21.065	16.47
1972	154.098	153.759	-.339	-.22
1973	144.262	155.558	11.296	7.83
1974	137.705	137.126	-.579	-.42

VAR. NO. 47 IS CMP

	ACTUAL	SOLN	DEV	P.DEV
1965	173.499	164.920	-8.579	-4.94
1966	203.490	186.418	-17.072	-8.39
1967	234.597	197.415	-37.182	-15.85
1968	244.718	213.086	-31.632	-12.93
1969	254.886	230.273	-24.613	-9.66
1970	264.239	231.532	-32.707	-12.38
1971	263.774	235.514	-28.260	-10.71
1972	262.611	238.567	-24.044	-9.16
1973	286.529	249.166	-37.363	-13.04
1974	260.982	229.400	-31.582	-12.10

Table 6–6 *(continued)*

THE WORLD ZINC INDUSTRY
DYNAMIC SIMULATION 1965-1974
MODEL ONE

(CONTINUED)

VAR. NO. 48 IS MMP

	ACTUAL	SOLN	DEV	P.DEV
1965	93.810	101.310	7.500	7.99
1966	91.396	100.237	8.841	9.67
1967	100.571	97.031	-3.540	-3.52
1968	100.088	98.695	-1.393	-1.39
1969	105.663	102.238	-3.425	-3.24
1970	111.106	103.849	-7.257	-6.53
1971	110.536	106.826	-3.710	-3.36
1972	113.345	109.401	-3.944	-3.48
1973	113.257	112.242	-1.015	-.90
1974	109.438	104.697	-4.741	-4.33

VAR. NO. 49 IS PMP

	ACTUAL	SOLN	DEV	P.DEV
1965	129.651	128.770	-.881	-.68
1966	144.665	123.289	-21.376	-14.78
1967	155.282	133.915	-21.367	-13.76
1968	148.418	139.721	-8.697	-5.86
1969	152.976	142.981	-9.995	-6.53
1970	152.332	149.016	-3.316	-2.18
1971	162.038	161.534	-.504	-.31
1972	182.145	174.161	-7.984	-4.38
1973	210.724	198.277	-12.447	-5.91
1974	197.158	185.493	-11.665	-5.92

VAR. NO. 50 IS EMP

	ACTUAL	SOLN	DEV	P.DEV
1965	96.190	98.114	1.924	2.00
1966	100.000	103.574	3.574	3.57
1967	105.714	107.793	2.079	1.97
1968	117.143	112.900	-4.243	-3.62
1969	128.571	119.513	-9.058	-7.05
1970	127.619	124.458	-3.161	-2.48
1971	123.810	123.770	-.040	-.03
1972	126.667	127.774	1.107	.87
1973	122.857	121.174	-1.683	-1.37
1974	120.000	118.950	-1.050	-.88

VAR. NO. 51 IS RWMP

	ACTUAL	SOLN	DEV	P.DEV
1965	115.482	116.265	.783	.68
1966	117.624	119.586	1.962	1.67
1967	116.748	119.560	2.812	2.41
1968	122.468	121.736	-.732	-.60
1969	128.213	125.247	-2.966	-2.31
1970	143.014	129.072	-13.942	-9.75
1971	152.483	135.700	-16.783	-11.01
1972	154.455	144.061	-10.394	-6.73
1973	157.765	160.614	2.849	1.81
1974	183.082	171.538	-11.544	-6.31

THE WORLD ZINC INDUSTRY
DYNAMIC SIMULATION 1965-1974
MODEL ONE

(CONTINUED)

VAR. NO. 53 IS UNS

	ACTUAL	SOLN	DEV	P.DEV
1965	131.940	134.120	2.180	1.65
1966	133.173	138.822	5.649	4.24
1967	116.452	121.683	5.231	4.49
1968	133.655	130.320	-3.335	-2.50
1969	143.301	129.716	-13.585	-9.48
1970	130.011	117.965	-12.046	-9.27
1971	135.691	123.837	-11.854	-8.74
1972	150.000	139.490	-10.510	-7.01
1973	143.033	147.686	4.653	3.25
1974	137.353	137.435	.082	.06

VAR. NO. 54 IS UOS

	ACTUAL	SOLN	DEV	P.DEV
1965	130.986	118.890	-12.096	-9.23
1966	137.852	121.612	-16.240	-11.78
1967	128.345	109.289	-19.056	-14.85
1968	127.465	115.131	-12.334	-9.68
1969	130.458	115.004	-15.454	-11.85
1970	115.141	107.499	-7.642	-6.64
1971	127.817	111.907	-15.910	-12.45
1972	126.761	122.787	-3.974	-3.14
1973	149.120	128.751	-20.369	-13.66
1974	140.845	124.307	-16.538	-11.74

VAR. NO. 55 IS RSCRAP

	ACTUAL	SOLN	DEV	P.DEV
1965	115.373	101.452	-13.921	-12.07
1966	111.557	102.834	-8.723	-7.82
1967	113.454	105.529	-7.925	-6.98
1968	122.356	110.762	-11.594	-9.48
1969	121.865	117.326	-4.539	-3.72
1970	133.066	118.181	-14.885	-11.19
1971	128.068	118.228	-9.840	-7.68
1972	133.556	125.426	-8.130	-6.09
1973	144.913	133.189	-11.724	-8.09
1974	136.992	130.506	-6.486	-4.73

VAR. NO. 57 IS UIMP

	ACTUAL	SOLN	DEV	P.DEV
1965	116.755	114.815	-1.940	-1.66
1966	165.074	146.163	-18.911	-11.46
1967	153.913	157.273	3.360	2.18
1968	167.379	163.439	-3.940	-2.35
1969	185.286	163.161	-22.125	-11.94
1970	160.154	159.473	-.681	-.43
1971	136.222	158.634	22.412	16.45
1972	166.371	171.397	5.026	3.02
1973	165.506	167.609	2.103	1.27
1974	160.418	158.750	-1.668	-1.04

Table 6–2 *(continued)*

THE WORLD ZINC INDUSTRY
DYNAMIC SIMULATION 1965-1974
MODEL ONE

(CONTINUED)

VAR. NO. 59 IS USTK

	ACTUAL	SOLN	DEV	P.DEV
1965	52.959	30.171	-22.788	-43.03
1966	102.663	107.364	4.701	4.58
1967	140.237	173.398	33.161	23.65
1968	111.538	122.407	10.869	9.74
1969	121.450	112.696	-8.754	-7.21
1970	209.615	178.242	-31.373	-14.97
1971	88.609	115.659	27.050	30.53
1972	60.355	39.168	-21.187	-35.10
1973	45.118	47.776	2.658	5.89
1974	118.343	116.487	-1.856	-1.57

VAR. NO. 60 IS RSTK

	ACTUAL	SOLN	DEV	P.DEV
1965	110.983	199.322	88.339	79.60
1966	123.187	61.416	-61.771	-50.14
1967	120.101	95.699	-24.402	-20.32
1968	108.040	129.080	21.040	19.47
1969	138.119	114.025	-24.094	-17.44
1970	178.536	152.973	-25.563	-14.32
1971	191.027	165.399	-25.628	-13.42
1972	145.154	196.238	51.084	35.19
1973	90.811	209.428	118.617	130.62
1974	477.387	396.325	-81.062	-16.98

VAR. NO. 171 IS USPZ

	ACTUAL	SOLN	DEV	P.DEV
1965	118.500	95.397	-23.103	-19.50
1966	114.028	103.941	-10.088	-8.85
1967	108.857	116.136	7.279	6.69
1968	104.180	102.576	-1.604	-1.54
1969	107.700	107.267	-.433	-.40
1970	108.199	119.823	11.624	10.74
1971	111.109	113.956	2.847	2.56
1972	116.547	117.971	1.423	1.22
1973	118.741	121.272	2.531	2.13
1974	174.177	169.660	-4.516	-2.59

VAR. NO. 172 IS LMPZ

	ACTUAL	SOLN	DEV	P.DEV
1965	137.925	89.246	-48.679	-35.29
1966	120.445	101.181	-19.263	-15.99
1967	98.205	111.135	12.930	13.17
1968	104.615	99.621	-4.994	-4.77
1969	111.268	112.705	1.437	1.29
1970	104.963	114.694	9.732	9.27
1971	105.366	123.986	18.620	17.67
1972	104.846	123.486	18.640	17.78
1973	218.406	199.268	-19.138	-8.76
1974	293.611	302.952	9.341	3.18

Table 6–3
Results of Dynamic Simulation: Model 2

THE WORLD ZINC INDUSTRY
DYNAMIC SIMULATION 1965-1974
MODEL TWO

VAR. NO. 1 IS RCN

	ACTUAL	SOLN	DEV	P.DEV
1965	109.545	118.222	8.677	7.92
1966	112.811	121.962	9.151	8.11
1967	119.096	124.294	5.198	4.36
1968	131.247	134.373	3.126	2.38
1969	146.669	145.496	-1.173	-.80
1970	148.066	152.867	4.801	3.24
1971	147.240	163.448	16.208	11.01
1972	164.650	172.499	7.849	4.77
1973	181.993	186.410	4.417	2.43
1974	172.762	176.556	3.794	2.20

VAR. NO. 2 IS UCN

	ACTUAL	SOLN	DEV	P.DEV
1965	122.099	115.725	-6.374	-5.22
1966	127.256	116.381	-10.875	-8.55
1967	111.750	114.443	2.693	2.41
1968	120.681	121.467	.786	.65
1969	119.567	120.474	.907	.76
1970	108.315	111.688	3.373	3.11
1971	113.453	118.291	4.838	4.26
1972	127.505	122.026	-5.479	-4.30
1973	134.963	127.409	-7.554	-5.60
1974	116.713	117.207	.494	.42

VAR. NO. 3 IS JCN

	ACTUAL	SOLN	DEV	P.DEV
1965	108.139	113.852	5.713	5.28
1966	127.568	122.034	-5.534	-4.34
1967	151.559	143.981	-7.578	-5.00
1968	171.546	167.893	-3.653	-2.13
1969	196.882	188.613	-8.269	-4.20
1970	204.496	210.147	5.651	2.76
1971	204.824	224.549	19.725	9.63
1972	235.215	237.692	2.477	1.05
1973	267.443	252.670	-14.773	-5.52
1974	222.678	231.456	8.778	3.94

VAR. NO. 4 IS KCN

	ACTUAL	SOLN	DEV	P.DEV
1965	102.082	104.914	2.832	2.77
1966	100.281	100.909	.628	.63
1967	95.527	97.420	1.893	1.98
1968	103.910	100.752	-3.158	-3.04
1969	107.539	105.165	-2.374	-2.21
1970	104.163	105.564	1.401	1.35
1971	100.731	105.799	5.068	5.03
1972	102.785	107.372	4.587	4.46
1973	111.111	107.015	-4.096	-3.69
1974	99.297	97.986	-1.311	-1.32

Table 6–3 *(continued)*

THE WORLD ZINC INDUSTRY
DYNAMIC SIMULATION 1965-1974
MODEL TWO

(CONTINUED)

VAR. NO. 5 IS GCN

	ACTUAL	SOLN	DEV	P.DEV
1965	116.967	120.347	3.380	2.89
1966	111.557	117.795	6.238	5.59
1967	104.809	105.886	1.077	1.03
1968	117.404	110.296	-7.108	-6.05
1969	101.940	118.325	16.385	16.07
1970	123.415	122.380	-1.035	-.84
1971	124.290	124.520	.230	.19
1972	127.295	129.585	2.290	1.80
1973	137.787	137.708	-.079	-.06
1974	121.503	127.611	6.108	5.03

VAR. NO. 6 IS FCN

	ACTUAL	SOLN	DEV	P.DEV
1965	105.876	106.350	.474	.45
1966	105.195	110.428	5.233	4.97
1967	108.277	109.962	1.685	1.56
1968	100.502	109.899	9.397	9.35
1969	112.218	113.237	1.019	.91
1970	108.026	112.391	4.365	4.04
1971	110.247	113.007	2.760	2.50
1972	124.686	117.637	-7.049	-5.65
1973	138.302	127.638	-10.664	-7.71
1974	133.286	125.136	-8.150	-6.11

VAR. NO. 7 IS R1CN

	ACTUAL	SOLN	DEV	P.DEV
1965	108.888	109.907	1.019	.94
1966	110.323	116.009	5.686	5.15
1967	116.687	118.337	1.650	1.41
1968	123.065	126.119	3.054	2.48
1969	139.087	134.584	-4.503	-3.24
1970	147.990	138.225	-9.765	-6.60
1971	133.903	145.602	11.699	8.74
1972	153.673	155.971	2.298	1.50
1973	168.833	168.285	-.548	-.32
1974	171.962	166.300	-5.662	-3.29

VAR. NO. 8 IS R2CN

	ACTUAL	SOLN	DEV	P.DEV
1965	110.614	112.672	2.058	1.86
1966	116.170	117.482	1.312	1.13
1967	125.700	124.284	-1.416	-1.13
1968	147.516	138.143	-9.373	-6.35
1969	156.594	153.013	-3.581	-2.29
1970	178.726	167.962	-10.764	-6.02
1971	188.347	197.370	9.023	4.79
1972	209.982	212.098	2.116	1.01
1973	231.030	247.433	16.403	7.10
1974	235.501	233.374	-2.127	-.90

THE WORLD ZINC INDUSTRY
DYNAMIC SIMULATION 1965-1974
MODEL TWO

(CONTINUED)

VAR. NO. 10 IS UCNG

	ACTUAL	SOLN	DEV	P.DEV
1965	114.765	111.332	-3.433	-2.99
1966	121.191	112.308	-8.883	-7.33
1967	112.405	114.464	2.059	1.83
1968	118.673	117.828	-.845	-.71
1969	117.388	119.979	2.591	2.21
1970	112.825	117.332	4.507	4.00
1971	112.956	120.472	7.516	6.65
1972	123.289	124.877	1.588	1.29
1973	134.146	125.247	-8.899	-6.63
1974	117.598	121.435	3.837	3.26

VAR. NO. 11 IS UCND

	ACTUAL	SOLN	DEV	P.DEV
1965	136.133	123.465	-12.668	-9.31
1966	129.334	118.891	-10.443	-8.07
1967	114.208	111.341	-2.867	-2.51
1968	120.136	122.784	2.648	2.20
1969	123.006	115.795	-7.211	-5.86
1970	98.941	99.156	.215	.22
1971	110.139	113.597	3.458	3.14
1972	123.712	114.262	-9.450	-7.64
1973	130.299	123.076	-7.223	-5.54
1974	91.861	105.248	13.387	14.57

VAR. NO. 12 IS UCNB

	ACTUAL	SOLN	DEV	P.DEV
1965	98.971	113.578	14.607	14.76
1966	144.340	121.668	-22.672	-15.71
1967	102.316	121.142	18.826	18.40
1968	125.986	125.708	-.278	-.22
1969	139.623	129.698	-9.925	-7.11
1970	99.400	125.947	26.547	26.71
1971	117.067	123.495	6.428	5.49
1972	149.485	130.571	-18.914	-12.65
1973	153.774	139.818	-13.956	-9.08
1974	137.736	136.122	-1.614	-1.17

VAR. NO. 13 IS UCNR

	ACTUAL	SOLN	DEV	P.DEV
1965	108.616	110.114	1.498	1.38
1966	124.543	109.788	-14.755	-11.85
1967	107.572	108.191	.619	.58
1968	115.927	108.650	-7.277	-6.28
1969	115.144	107.968	-7.176	-6.23
1970	97.128	104.896	7.768	8.00
1971	91.906	104.297	12.391	13.48
1972	107.050	104.520	-2.530	-2.36
1973	96.606	102.683	6.077	6.29
1974	91.645	101.006	9.361	10.21

Table 6–3 *(continued)*

THE WORLD ZINC INDUSTRY
DYNAMIC SIMULATION 1965-1974
MODEL TWO

(CONTINUED)

VAR. NO. 14 IS UCNO

	ACTUAL	SOLN	DEV	P.DEV
1965	123.487	122.001	-1.486	-1.20
1966	131.179	131.662	.483	.37
1967	121.641	133.254	11.613	9.55
1968	135.795	138.784	2.989	2.20
1969	149.744	144.255	-5.489	-3.67
1970	143.897	138.784	-5.113	-3.55
1971	134.872	138.784	3.912	2.90
1972	149.949	147.356	-2.593	-1.73
1973	162.256	158.833	-3.423	-2.11
1974	164.513	158.075	-6.438	-3.91

VAR. NO. 15 IS UCNM

	ACTUAL	SOLN	DEV	P.DEV
1965	118.148	50.532	-67.616	-57.23
1966	139.630	65.789	-73.841	-52.88
1967	122.222	75.052	-47.170	-38.59
1968	144.815	89.495	-55.320	-38.20
1969	154.815	93.192	-61.623	-39.80
1970	122.222	79.549	-42.673	-34.91
1971	113.704	84.849	-28.855	-25.38
1972	104.074	100.485	-3.589	-3.45
1973	98.519	94.246	-4.273	-4.34
1974	28.889	39.612	10.723	37.12

VAR. NO. 17 IS JCNG

	ACTUAL	SOLN	DEV	P.DEV
1965	119.628	116.270	-3.358	-2.81
1966	125.145	121.311	-3.834	-3.06
1967	151.452	141.595	-9.857	-6.51
1968	167.073	165.150	-1.923	-1.15
1969	190.186	186.184	-4.002	-2.10
1970	195.877	208.754	12.877	6.57
1971	196.516	224.387	27.871	14.18
1972	221.835	236.092	14.257	6.43
1973	263.298	238.410	-24.888	-9.45
1974	215.273	209.211	-6.062	-2.82

VAR. NO. 18 IS JCND

	ACTUAL	SOLN	DEV	P.DEV
1965	80.110	99.602	19.492	24.33
1966	112.707	115.681	2.974	2.64
1967	160.773	149.887	-10.886	-6.77
1968	191.897	182.363	-9.534	-4.97
1969	229.834	201.645	-28.189	-12.26
1970	226.703	220.774	-5.929	-2.62
1971	235.543	243.993	8.450	3.59
1972	276.796	267.837	-8.959	-3.24
1973	291.160	300.872	9.712	3.34
1974	256.169	281.422	25.253	9.86

THE WORLD ZINC INDUSTRY
DYNAMIC SIMULATION 1965-1974
MODEL TWO

(CONTINUED)

VAR. NO. 19 IS JCNR

	ACTUAL	SOLN	DEV	P.DEV
1965	108.313	118.375	10.062	9.29
1966	127.628	128.101	.473	.37
1967	147.433	143.573	-3.860	-2.62
1968	160.880	158.927	-1.953	-1.21
1969	178.973	175.536	-3.437	-1.92
1970	188.020	192.640	4.620	2.46
1971	186.797	198.508	11.711	6.27
1972	219.804	205.145	-14.659	-6.67
1973	255.990	226.844	-29.146	-11.39
1974	182.396	222.612	40.216	22.05

VAR. NO. 20 IS JCNR

	ACTUAL	SOLN	DEV	P.DEV
1965	118.954	140.699	21.745	18.28
1966	162.092	145.224	-16.868	-10.41
1967	159.477	158.775	-.702	-.44
1968	190.850	176.998	-13.852	-7.26
1969	206.536	194.664	-11.872	-5.75
1970	232.680	203.747	-28.933	-12.43
1971	175.817	196.004	20.187	11.48
1972	208.497	203.676	-4.821	-2.31
1973	256.209	226.193	-30.016	-11.72
1974	166.667	228.730	62.063	37.24

VAR. NO. 21 IS JCNO

	ACTUAL	SOLN	DEV	P.DEV
1965	75.556	93.274	17.718	23.45
1966	93.333	98.758	5.425	5.81
1967	99.259	107.023	7.764	7.82
1968	115.556	114.925	-.631	-.55
1969	118.519	123.582	5.063	4.27
1970	146.667	132.108	-14.559	-9.93
1971	136.296	136.620	.324	.24
1972	140.741	140.071	-.670	-.48
1973	155.556	149.695	-5.861	-3.77
1974	142.222	147.567	5.345	3.76

VAR. NO. 22 IS JCNM

	ACTUAL	SOLN	DEV	P.DEV
1965	84.706	118.564	33.858	39.97
1966	262.353	143.313	-119.040	-45.37
1967	181.176	188.600	7.424	4.10
1968	238.824	241.817	2.993	1.25
1969	310.588	309.784	-.804	-.26
1970	357.647	390.090	32.443	9.07
1971	424.706	419.790	-4.916	-1.16
1972	512.941	450.234	-62.707	-12.23
1973	451.765	568.827	117.062	25.91
1974	580.000	543.338	-36.662	-6.32

Table 6–3 *(continued)*

THE WORLD ZINC INDUSTRY
DYNAMIC SIMULATION 1965-1974
MODEL TWO

(CONTINUED)

VAR. NO. 24 IS KCNG

	ACTUAL	SOLN	DEV	P.DEV
1965	107.489	114.825	7.336	6.82
1966	105.947	111.211	5.264	4.97
1967	105.286	105.177	-.109	-.10
1968	105.617	102.610	-3.007	-2.85
1969	109.031	104.246	-4.785	-4.39
1970	106.167	105.066	-1.101	-1.04
1971	108.590	104.935	-3.655	-3.37
1972	110.132	105.009	-5.123	-4.65
1973	112.445	104.568	-7.877	-7.00
1974	101.652	95.184	-6.468	-6.36

VAR. NO. 25 IS KCND

	ACTUAL	SOLN	DEV	P.DEV
1965	106.038	106.630	.592	.56
1966	101.031	103.281	2.250	2.23
1967	95.876	99.652	3.776	3.94
1968	106.480	107.600	1.120	1.05
1969	114.433	108.514	-5.919	-5.17
1970	110.015	107.312	-2.703	-2.46
1971	102.209	109.834	7.625	7.46
1972	107.658	113.402	5.744	5.34
1973	117.231	113.168	-4.063	-3.47
1974	102.798	102.844	.046	.04

VAR. NO. 26 IS KCNB

	ACTUAL	SOLN	DEV	P.DEV
1965	100.000	93.315	-6.685	-6.69
1966	92.176	86.153	-6.023	-6.53
1967	82.559	82.594	.035	.04
1968	92.991	87.777	-5.214	-5.61
1969	94.866	96.816	1.950	2.06
1970	87.531	96.995	9.464	10.81
1971	81.418	95.990	14.572	17.90
1972	81.500	95.658	14.158	17.37
1973	91.769	94.122	2.353	2.56
1974	81.989	82.193	.204	.25

VAR. NO. 27 IS KCNR

	ACTUAL	SOLN	DEV	P.DEV
1965	107.265	111.003	3.738	3.49
1966	116.239	107.646	-8.593	-7.39
1967	99.145	104.716	5.571	5.62
1968	110.256	106.915	-3.341	-3.03
1969	116.239	111.177	-5.062	-4.35
1970	110.256	112.656	2.400	2.18
1971	112.821	113.573	.752	.67
1972	107.692	114.696	7.004	6.50
1973	123.504	116.071	-7.433	-6.02
1974	94.872	106.692	11.820	12.46

THE WORLD ZINC INDUSTRY
DYNAMIC SIMULATION 1965-1974
MODEL TWO

(CONTINUED)

VAR. NO. 28 IS KCNO

	ACTUAL	SOLN	DEV	P.DEV
1965	100.000	117.621	17.621	17.62
1966	109.091	120.037	10.946	10.03
1967	120.553	122.384	1.831	1.52
1968	133.597	126.155	-7.442	-5.57
1969	136.759	129.413	-7.346	-5.37
1970	143.083	132.556	-10.527	-7.36
1971	139.526	133.580	-5.946	-4.26
1972	143.083	145.367	2.284	1.60
1973	167.194	142.630	-24.564	-14.69
1974	150.198	145.667	-4.531	-3.02

VAR. NO. 29 IS KCNM

	ACTUAL	SOLN	DEV	P.DEV
1965	95.669	102.651	6.982	7.30
1966	93.701	103.739	10.038	10.71
1967	94.094	103.739	9.645	10.25
1968	107.087	107.489	.402	.38
1969	108.661	110.114	1.453	1.34
1970	117.717	110.633	-7.084	-6.02
1971	114.173	110.633	-3.540	-3.10
1972	121.654	111.668	-9.986	-8.21
1973	116.142	117.746	1.604	1.38
1974	117.323	115.744	-1.579	-1.35

VAR. NO. 30 IS GCNG

	ACTUAL	SOLN	DEV	P.DEV
1965	112.955	125.954	12.999	11.51
1966	123.448	127.857	4.409	3.57
1967	125.803	121.617	-4.186	-3.33
1968	138.009	134.368	-3.641	-2.64
1969	149.251	148.758	-.493	-.33
1970	147.537	155.330	7.793	5.28
1971	151.178	151.323	.145	.10
1972	165.418	156.589	-8.829	-5.34
1973	171.306	159.619	-11.687	-6.82
1974	158.137	148.406	-9.731	-6.15

VAR. NO. 31 IS GCND

	ACTUAL	SOLN	DEV	P.DEV
1965	127.553	132.297	4.744	3.72
1966	124.941	136.529	11.588	9.27
1967	137.530	115.884	-21.646	-15.74
1968	165.796	145.364	-20.432	-12.32
1969	185.511	174.345	-11.166	-6.02
1970	197.862	177.186	-20.676	-10.45
1971	183.848	179.140	-4.708	-2.56
1972	177.672	178.145	.473	.27
1973	193.112	194.099	.987	.51
1974	146.081	163.417	17.336	11.87

Table 6–3 *(continued)*

THE WORLD ZINC INDUSTRY
DYNAMIC SIMULATION 1965-1974
MODEL TWO

(CONTINUED)

VAR. NO. 32 IS GCNB

	ACTUAL	SOLN	DEV	P.DEV
1965	123.534	121.877	-1.657	-1.34
1966	107.115	120.420	13.305	12.42
1967	98.593	106.318	7.725	7.83
1968	108.991	100.125	-8.866	-8.14
1969	120.172	99.491	-20.681	-17.21
1970	111.572	103.471	-8.101	-7.26
1971	100.547	110.642	10.095	10.04
1972	107.428	118.116	10.688	9.95
1973	124.941	128.858	3.917	3.13
1974	110.399	128.940	18.541	16.79

VAR. NO. 33 IS GCNR

	ACTUAL	SOLN	DEV	P.DEV
1965	111.930	106.208	-5.722	-5.11
1966	100.134	93.121	-7.013	-7.00
1967	76.005	81.327	5.322	7.00
1968	84.048	78.016	-6.032	-7.18
1969	92.895	82.875	-10.020	-10.79
1970	78.284	87.630	9.346	11.94
1971	91.287	90.153	-1.134	-1.24
1972	83.780	91.775	7.995	9.54
1973	84.718	89.724	5.006	5.91
1974	78.150	74.609	-3.541	-4.53

VAR. NO. 34 IS GCNO

	ACTUAL	SOLN	DEV	P.DEV
1965	103.665	122.936	19.271	18.59
1966	113.613	133.418	19.805	17.43
1967	128.272	139.420	11.148	8.69
1968	147.120	155.166	8.046	5.47
1969	165.969	160.003	-5.966	-3.59
1970	143.455	149.737	6.282	4.38
1971	171.204	140.404	-30.800	-17.99
1972	160.209	143.439	-16.770	-10.47
1973	171.204	168.109	-3.095	-1.81
1974	155.497	177.003	21.506	13.83

VAR. NO. 35 IS GCNM

	ACTUAL	SOLN	DEV	P.DEV
1965	85.393	95.928	10.535	12.34
1966	78.652	59.085	-19.567	-24.88
1967	10.112	21.074	10.962	108.40
1968	8.989	12.146	3.157	35.12
1969	10.112	12.265	2.153	21.29
1970	23.596	21.610	-1.986	-8.42
1971	77.528	38.270	-39.258	-50.64
1972	68.539	68.502	-.037	-.05
1973	82.022	105.143	23.121	28.19
1974	74.157	59.049	-15.108	-20.37

THE WORLD ZINC INDUSTRY
DYNAMIC SIMULATION 1965-1974
MODEL TWO

(CONTINUED)

VAR. NO. 37 IS FCNG

	ACTUAL	SOLN	DEV	P.DEV
1965	101.977	109.394	7.417	7.27
1966	98.870	114.559	15.689	15.87
1967	107.486	115.728	8.242	7.67
1968	97.316	115.674	18.358	18.86
1969	112.147	121.256	9.109	8.12
1970	107.627	123.106	15.479	14.38
1971	114.407	121.450	7.043	6.16
1972	132.910	123.388	-9.522	-7.16
1973	139.831	126.676	-13.155	-9.41
1974	141.243	120.322	-20.921	-14.81

VAR. NO. 38 IS FCND

	ACTUAL	SOLN	DEV	P.DEV
1965	98.119	90.777	-7.342	-7.48
1966	106.270	105.828	-.442	-.42
1967	98.119	109.230	11.111	11.32
1968	105.956	110.328	4.372	4.13
1969	130.094	112.785	-17.309	-13.30
1970	112.539	110.990	-1.549	-1.38
1971	106.897	130.377	23.480	21.97
1972	178.683	156.833	-21.850	-12.23
1973	175.862	182.062	6.200	3.53
1974	172.100	167.120	-4.980	-2.89

VAR. NO. 39 IS FCNB

	ACTUAL	SOLN	DEV	P.DEV
1965	113.636	72.012	-41.624	-36.63
1966	113.636	70.829	-42.807	-37.67
1967	95.455	66.867	-28.588	-29.95
1968	104.545	68.353	-36.192	-34.62
1969	106.818	75.259	-31.559	-29.54
1970	134.091	78.152	-55.939	-41.72
1971	104.545	79.802	-24.743	-23.67
1972	131.818	80.576	-51.242	-38.87
1973	136.364	91.491	-44.873	-32.91
1974	134.091	92.538	-41.553	-30.99

VAR. NO. 40 IS FCNR

	ACTUAL	SOLN	DEV	P.DEV
1965	111.194	109.720	-1.474	-1.33
1966	113.433	112.635	-.798	-.70
1967	114.055	112.525	-1.530	-1.34
1968	103.731	109.929	6.198	5.98
1969	110.821	106.045	-4.776	-4.31
1970	101.866	102.892	1.026	1.01
1971	104.478	103.693	-.785	-.75
1972	104.851	108.129	3.278	3.13
1973	115.423	110.759	-4.664	-4.04
1974	106.219	109.664	3.445	3.24

Table 6–3 *(continued)*

THE WORLD ZINC INDUSTRY
DYNAMIC SIMULATION 1965-1974
MODEL TWO

(CONTINUED)

VAR. NO. 41 IS FCNO

	ACTUAL	SOLN	DEV	P.DEV
1965	92.060	101.153	9.093	9.88
1966	101.717	109.086	7.369	7.25
1967	107.296	113.305	6.009	5.60
1968	106.009	116.307	10.298	9.71
1969	110.515	119.295	8.780	7.94
1970	100.858	113.538	12.680	12.57
1971	103.004	106.213	3.209	3.12
1972	96.567	106.978	10.411	10.78
1973	135.193	117.717	-17.476	-12.93
1974	135.408	122.384	-13.024	-9.62

VAR. NO. 42 IS FCNM

	ACTUAL	SOLN	DEV	P.DEV
1965	121.556	115.201	-6.355	-5.23
1966	102.444	108.398	5.954	5.81
1967	108.667	97.485	-11.182	-10.29
1968	88.889	97.784	8.895	10.01
1969	104.444	111.123	6.679	6.39
1970	121.333	115.546	-5.787	-4.77
1971	124.444	114.231	-10.213	-8.21
1972	100.000	112.368	12.368	12.37
1973	140.000	134.446	-5.554	-3.97
1974	140.222	136.528	-3.694	-2.63

VAR. NO. 44 IS RMP

	ACTUAL	SOLN	DEV	P.DEV
1965	118.979	120.195	1.216	1.02
1966	128.582	126.660	-1.922	-1.49
1967	139.412	131.154	-8.258	-5.92
1968	145.556	137.944	-7.612	-5.23
1969	156.276	145.552	-10.724	-6.86
1970	160.459	149.645	-10.814	-6.74
1971	160.876	154.365	-6.511	-4.05
1972	167.687	159.767	-7.920	-4.72
1973	173.418	166.283	-7.135	-4.11
1974	169.831	159.136	-10.695	-6.30

VAR. NO. 45 IS UMP

	ACTUAL	SOLN	DEV	P.DEV
1965	115.476	112.613	-2.863	-2.48
1966	108.186	110.939	2.753	2.54
1967	103.812	106.329	2.517	2.42
1968	100.042	99.928	-.114	-.11
1969	104.520	98.685	-5.835	-5.58
1970	100.937	96.737	-4.200	-4.16
1971	94.959	96.100	1.141	1.20
1972	90.377	96.845	6.468	7.16
1973	90.481	86.520	-3.961	-4.38
1974	93.439	91.596	-1.843	-1.97

THE WORLD ZINC INDUSTRY
DYNAMIC SIMULATION 1965-1974
MODEL TWO

(CONTINUED)

VAR. NO. 46 IS AMP

	ACTUAL	SOLN	DEV	P.DEV
1965	98.361	109.324	10.963	11.15
1966	104.918	113.510	8.592	8.19
1967	113.115	117.402	4.287	3.79
1968	118.033	125.217	7.184	6.09
1969	142.623	132.259	-10.364	-7.27
1970	136.066	139.906	3.840	2.82
1971	127.869	148.934	21.065	16.47
1972	154.098	153.759	-.339	-.22
1973	144.262	155.558	11.296	7.83
1974	137.705	137.126	-.579	-.42

VAR. NO. 47 IS CMP

	ACTUAL	SOLN	DEV	P.DEV
1965	173.499	164.920	-8.579	-4.94
1966	203.490	186.418	-17.072	-8.39
1967	234.597	197.415	-37.182	-15.85
1968	244.718	213.086	-31.632	-12.93
1969	254.886	230.273	-24.613	-9.66
1970	264.239	231.532	-32.707	-12.38
1971	263.774	235.514	-28.260	-10.71
1972	262.611	238.567	-24.044	-9.16
1973	286.529	249.166	-37.363	-13.04
1974	260.982	229.400	-31.582	-12.10

VAR. NO. 48 IS MMP

	ACTUAL	SOLN	DEV	P.DEV
1965	93.810	101.310	7.500	7.99
1966	91.396	100.237	8.841	9.67
1967	100.571	97.031	-3.540	-3.52
1968	100.088	98.695	-1.393	-1.39
1969	105.663	102.238	-3.425	-3.24
1970	111.106	103.849	-7.257	-6.53
1971	110.536	106.826	-3.710	-3.36
1972	113.345	109.401	-3.944	-3.48
1973	113.257	112.242	-1.015	-.90
1974	109.438	104.697	-4.741	-4.33

VAR. NO. 49 IS PMP

	ACTUAL	SOLN	DEV	P.DEV
1965	129.651	128.770	-.881	-.68
1966	144.665	123.289	-21.376	-14.78
1967	155.282	133.915	-21.367	-13.76
1968	148.418	139.721	-8.697	-5.86
1969	152.976	142.981	-9.995	-6.53
1970	152.332	149.016	-3.316	-2.18
1971	162.038	161.534	-.504	-.31
1972	182.145	174.161	-7.984	-4.38
1973	210.724	198.277	-12.447	-5.91
1974	197.158	185.493	-11.665	-5.92

Table 6–3 *(continued)*

THE WORLD ZINC INDUSTRY
DYNAMIC SIMULATION 1965-1974
MODEL TWO

(CONTINUED)

VAR. NO. 50 IS EMP

	ACTUAL	SOLN	DEV	P.DEV
1965	96.190	98.114	1.924	2.00
1966	100.000	103.574	3.574	3.57
1967	105.714	107.793	2.079	1.97
1968	117.143	112.900	-4.243	-3.62
1969	128.571	119.513	-9.058	-7.05
1970	127.619	124.458	-3.161	-2.48
1971	123.810	123.770	-.040	-.03
1972	126.667	127.774	1.107	.87
1973	122.857	121.174	-1.683	-1.37
1974	120.000	118.950	-1.050	-.88

VAR. NO. 51 IS RWMP

	ACTUAL	SOLN	DEV	P.DEV
1965	115.482	116.265	.783	.68
1966	117.624	119.586	1.962	1.67
1967	116.748	119.560	2.812	2.41
1968	122.468	121.736	-.732	-.60
1969	128.213	125.247	-2.966	-2.31
1970	143.014	129.072	-13.942	-9.75
1971	152.483	135.700	-16.783	-11.01
1972	154.455	144.061	-10.394	-6.73
1973	157.765	160.614	2.849	1.81
1974	183.082	171.538	-11.544	-6.31

VAR. NO. 53 IS UNS

	ACTUAL	SOLN	DEV	P.DEV
1965	131.940	134.120	2.180	1.65
1966	133.173	138.822	5.649	4.24
1967	116.452	121.683	5.231	4.49
1968	133.655	130.320	-3.335	-2.50
1969	143.301	129.716	-13.585	-9.48
1970	130.011	117.965	-12.046	-9.27
1971	135.691	123.837	-11.854	-8.74
1972	150.000	139.490	-10.510	-7.01
1973	143.033	147.686	4.653	3.25
1974	137.353	137.435	.082	.06

VAR. NO. 54 IS UOS

	ACTUAL	SOLN	DEV	P.DEV
1965	130.986	118.890	-12.096	-9.23
1966	137.852	121.612	-16.240	-11.78
1967	128.345	109.289	-19.056	-14.85
1968	127.465	115.131	-12.334	-9.68
1969	130.458	115.004	-15.454	-11.85
1970	115.141	107.499	-7.642	-6.64
1971	127.817	111.907	-15.910	-12.45
1972	126.761	122.787	-3.974	-3.14
1973	149.120	128.751	-20.369	-13.66
1974	140.845	124.307	-16.538	-11.74

THE WORLD ZINC INDUSTRY
DYNAMIC SIMULATION 1965-1974
MODEL TWO

(CONTINUED)

VAR. NO. 55 IS RSCRAP

	ACTUAL	SOLN	DEV	P.DEV
1965	115.373	101.452	-13.921	-12.07
1966	111.557	102.834	-8.723	-7.82
1967	113.454	105.529	-7.925	-6.98
1968	122.356	110.762	-11.594	-9.48
1969	121.865	117.326	-4.539	-3.72
1970	133.066	118.181	-14.885	-11.19
1971	128.068	118.228	-9.840	-7.68
1972	133.556	125.426	-8.130	-6.09
1973	144.913	133.189	-11.724	-8.09
1974	136.992	130.506	-6.486	-4.73

VAR. NO. 57 IS UIMP

	ACTUAL	SOLN	DEV	P.DEV
1965	116.755	114.756	-1.999	-1.71
1966	165.074	146.174	-18.900	-11.45
1967	153.913	157.282	3.369	2.19
1968	167.379	163.447	-3.932	-2.35
1969	185.286	163.117	-22.169	-11.96
1970	160.154	159.403	-.751	-.47
1971	136.222	158.608	22.386	16.43
1972	166.371	171.396	5.025	3.02
1973	165.506	167.586	2.080	1.26
1974	160.418	158.667	-1.751	-1.09

VAR. NO. 59 IS USTK

	ACTUAL	SOLN	DEV	P.DEV
1965	52.959	30.171	-22.788	-43.03
1966	102.663	107.364	4.701	4.58
1967	140.237	173.398	33.161	23.65
1968	111.538	122.407	10.869	9.74
1969	121.450	112.696	-8.754	-7.21
1970	209.615	178.242	-31.373	-14.97
1971	88.609	115.659	27.050	30.53
1972	60.355	39.168	-21.187	-35.10
1973	45.118	47.776	2.658	5.89
1974	118.343	116.487	-1.856	-1.57

VAR. NO. 60 IS RSTK

	ACTUAL	SOLN	DEV	P.DEV
1965	110.983	199.322	88.339	79.60
1966	123.187	62.821	-60.366	-49.00
1967	120.101	95.341	-24.760	-20.62
1968	108.040	128.741	20.701	19.16
1969	138.119	113.633	-24.486	-17.73
1970	178.536	154.032	-24.504	-13.72
1971	191.027	167.332	-23.695	-12.40
1972	145.154	196.552	51.398	35.41
1973	90.811	208.994	118.183	130.14
1974	477.387	397.131	-80.256	-16.81

Table 6–3 *(continued)*

THE WORLD ZINC INDUSTRY
DYNAMIC SIMULATION 1965-1974
MODEL TWO

(CONTINUED)

VAR. NO. 171 IS USPZ

	ACTUAL	SOLN	DEV	P.DEV
1965	118.500	95.635	-22.865	-19.30
1966	114.028	103.863	-10.165	-8.91
1967	108.857	116.056	7.199	6.61
1968	104.180	102.462	-1.718	-1.65
1969	107.700	107.406	-.294	-.27
1970	108.199	120.100	11.901	11.00
1971	111.109	113.900	2.790	2.51
1972	116.547	117.789	1.242	1.07
1973	118.741	121.367	2.627	2.21
1974	174.177	170.865	-3.311	-1.90

VAR. NO. 172 IS LMPZ

	ACTUAL	SOLN	DEV	P.DEV
1965	137.925	89.834	-48.091	-34.87
1966	120.445	101.032	-19.413	-16.12
1967	98.205	110.993	12.788	13.02
1968	104.615	99.457	-5.158	-4.93
1969	111.268	113.149	1.880	1.69
1970	104.963	115.503	10.541	10.04
1971	105.366	124.117	18.751	17.80
1972	104.846	123.304	18.459	17.61
1973	218.406	199.605	-18.801	-8.61
1974	293.611	306.017	12.406	4.23

Notes

1. See Friedman (1952, p. 456) "The only relevant test of a hypothesis is comparison of its predictions with what occurs: the hypothesis is rejected if its predictions are contradicted ("frequently" or more often than predictions from an alternative hypothesis); it is accepted if its predictions are not contradicted; great confidence is attached to it if it has survived many opportunities for contradiction."

2. The solution program uses a Gauss-Seidel iterative technique designed for solving intertemporal, nonlinear econometric models.

7 Dynamic Multiplier Simulations with the Econometric Models

Analytical solutions with simultaneous nonlinear dynamic models of a commodity market, such as the one under discussion, are extremely difficult. An alternative computer methodology that has recently been developed describes the system's response to exogenous shocks to the estimated econometric models. The so-called methodology of simulation enables the researcher to study the stability properties and policy implications of the models.[1]

Dynamic solutions of the models, which include the effects of exogenous changes in one or more variables or parameters of the system, when compared to the base solutions (also called control solutions), provide measures of the response of the model (also called dynamic multiplier solutions) to the exogenous changes. For six such changes imposed upon the system, dynamic multiplier solutions have been calculated. The control solutions used in this chapter are the same as those given in the last chapter for both models.

Multiplier Simulation Experiments

The set of six exogenous changes used to study the response properties of the models of the zinc industry include changes in economic activity in the major consumer countries, a change in technology in one of the most important consumer industries, a change in the price of the major substitute for zinc, and changes in U.S. government policy regarding its strategic stockpile program. These changes are classified in a set of four assumptions, thus constituting four experiments, depending on the nature of the variables involved, as follows:

1. (a). A one-time (1965) increase of 1 percent in the economic activity of the major zinc consumer countries (the United States, Japan, the United Kingdom, West Germany, and France)
   (b). A continued 1-percent increase each year (1965–1974) in the economic activity of the major zinc consumer countries.
2. A one-third decline in consumption requirements of zinc by the automobile industry throughout the world.
3. A one-time (1965) increase in the price of aluminum by 10 percent.

4. (a). A one-time (1965) increase in the U.S. strategic stockpile by 1 percent.
   (b). A continued decrease in the U.S. strategic stockpile by 100,000 tons per year, beginning in 1963.

Assumption 1(a) may be viewed as a sudden but temporary improvement in the economic activity of the large industrial countries. In 1(b), the same increase in economic activity continues throughout the simulation period. That is, in each year of simulation, the level of economic activity is 1 percent higher than in the base solution. Whereas in 1(a) an attempt is made to study the stability of the system to a temporary shock, in 1(b), the system's response to a persistent change is evaluated.

In the second experiment the possible effect of some technological change on the performance of the different constituents of the market is examined. In particular, what is being sought is the response of the system to the often-repeated possibility of a larger proportion of smaller or light-weight cars, or a development of a municipal mass transit system, or some other such development that would reduce consumption requirements of zinc in one of its major uses. This change is imposed on the system through reducing the coefficient of the activity variable in the consumption for die-casting equations (Model 2) in all the major consumer countries.

The third simulation assumption focuses on the likely changes in the behavior and performance of the zinc industry in response to an increase in the price of aluminum, the major substitute for zinc, by the International Bauxite Association. An increase in the price of aluminum raises the competitive strength of zinc vis-à-vis aluminum and other substitute materials such as plastics. The question posed is whether the zinc industry, given the substitution structure of zinc vis-à-vis other materials, will have any substantial gains (or losses, for a decrease in price, assuming the substitution effects are symmetric) in the long run.

The fourth simulation assumption attempts to evaluate the likely dynamic effects of the changes in the stockpile policy of the U.S. government. The major aims of the stockpile policy throughout the period have been protection of the domestic industry and possibly stabilization of the world zinc market.[2] In 4(a) an attempt is made to test the validity of these objectives through a temporary increase (1 percent) in the stockpile by the U.S. government. In 4(b), a similar question is posed in a very different form. Would the zinc industry have been unstable if the U.S. government had not intervened in the performance of the industry? In fact, this assumption serves a dual purpose. It is possible to think of the whole U.S. stockpile in 1963 (when it stood at about 1.4 million tons, more than half the world production of zinc in that year), as a new mine discovery and trace the effect of such a change on the price of zinc. Thus, in this multiplier simulation, the U.S. government is expected not to intervene in the world zinc market, but to decumulate the stocks evenly over the next fifteen years.

In the next section the impact of these multiplier simulations is discussed. Tables 7–1 and 7–4 contain the results of these multiplier simulations for major variables. The results are reported in terms of ratios of multiplier solutions to control solutions, multiplied by 100.

Results of Multiplier Simulation Experiments

Experiment 1

The multiplier simulation results of this experiment for 1(a) and 1(b) are given in table 7–1. In the first year, U.S. producers are attempting to stabilize the likely increase in price, but seem to have overdone this. The rest-of-the-world producers increase their stocks marginally for transactions and probably for speculative purposes. The result is a temporary fall in both prices. In the second year, both prices show a marginal improvement and, by the fourth year, a return to a base solution. The number of years taken to stabilize the price indicates delayed responses. In 1(b), the results are similar. The shift in demand produces a higher price between 0.1 and 0.3 above the base simulation. Long lags in both demand and supply do not seem to have allowed full adjustment by the end of the simulation period. In general, the market seems to be stable except for the differential lags in demand and supply resulting in minor fluctuations.

Experiment 2

Consumption of zinc, as reported in table 7–2, seems to be very sensitive to changes in the activity coefficient in the die-casting sector. The model reveals a substantial response to a development such as a change in automobile technology.

The consumption response, as expected, differs from one country to another depending on the technological requirements of zinc, different lags involved, and other considerations.[3] Since the model involves dynamic interaction of consumption and prices, the fall in consumption is substantial. In most countries, the fall in consumption is between 10 to 30 percent as compared to the control solution. Fall in demand in relation to supply results

Table 7–1
Multiplier Simulations: An Increase in the Economic Activity in the World (Model 1)

	Assumption 1a.					*Assumption 1b.*				
Year	*USTK*	*RSTK*	*USPZ*	*LMPZ*	*UIMP*	*USTK*	*RSTK*	*USPZ*	*LMPZ*	*UIMP*
1965	76.41	100.53	99.74	99.32	100.59	76.45	100.00	100.20	100.52	100.50
1966	100.00	97.65	100.28	100.12	100.02	93.28	101.82	100.25	100.05	100.45
1967	100.00	100.29	99.99	99.98	100.00	95.92	100.12	100.30	100.18	100.40
1968	100.00	99.96	100.00	100.00	100.00	93.68	100.35	100.37	100.24	100.42
1969	100.00	100.00	100.00	100.00	100.00	92.98	100.51	100.34	100.19	100.43
1970	100.00	100.00	100.00	100.00	100.00	95.89	100.33	100.33	100.25	100.41
1971	100.00	100.00	100.00	100.00	100.00	93.07	100.41	100.36	100.24	100.45
1972	100.00	100.00	100.00	100.00	100.00	97.93	100.37	100.36	100.28	100.45
1973	100.00	100.00	100.00	100.00	100.00	80.73	100.40	100.35	100.19	100.50
1974	100.00	100.00	100.00	100.00	100.00	92.69	100.22	100.32	100.21	100.49

Note: Data are (Multiplier Solution/Control Solution) × 100.0.

Table 7–2
Multiplier Simulations: A Technological Change in the Automobile Sector (Model 2)

Year	*UCN*	*JCN*	*KCN*	*GCN*	*FCN*	*R1CN*	*R2CN*	*USPZ*	*LMPZ*	*UIMP*
1965	70.17	89.58	88.21	89.91	93.34	56.40	73.41	81.72	52.00	104.63
1966	71.63	88.50	88.18	89.34	92.38	56.05	73.14	106.33	117.11	98.95
1967	74.78	87.02	88.23	90.04	92.10	55.91	72.19	87.22	70.23	101.84
1968	74.05	86.16	87.55	87.82	91.98	55.51	72.32	94.87	91.96	100.28
1969	76.57	86.21	87.96	88.21	91.91	55.11	71.95	91.64	83.45	100.90
1970	79.49	86.31	88.17	86.41	91.91	54.94	71.64	91.77	81.31	101.07
1971	77.32	85.73	87.87	86.47	90.47	54.69	71.17	90.24	81.13	101.08
1972	77.91	85.12	87.60	87.08	88.90	54.38	70.97	89.16	76.71	101.33
1973	77.46	84.12	87.59	86.70	88.04	53.97	70.39	89.45	84.62	100.82
1974	79.68	83.86	87.83	88.04	88.82	53.97	70.28	84.50	79.00	101.14

Note: Data are (Multiplier Solution/Control Solution) × 100.0.

in a fall in both the LME and the U.S. prices. However, as the free-market price is more senitive to changes in demand and supply forces, the fall in the LME price is larger than the U.S. price. This is also reflected in a small rise in imports into the United States.

Experiment 3

In the third experiment, the price of aluminum, the major substitute for zinc, was increased by 10 percent in 1965. Assuming substitution effects to be symmetric, results can easily be reinterpreted for a fall in the price of substitutes and the resultant effects on the price of zinc. As shown in table 7–3, in general, the effects of an increase in the price of aluminum are to increase the consumption of zinc, and, as a consequence, also to raise the price of zinc. The increase in price is, however, very small and reflects the lags involved in the system. At the end of the simulation period, the system becomes stabilized to the control solution, though the number of years taken differs with different countries and variables, according to the different structures of lags involved.

Experiment 4

In simulation 4(a), as may be seen in table 7–4, the effect of a 1-percent increase in the stockpile in 1965 continues up to the end of the fourth year. Further, price, rather than showing a steady change, fluctuates widely as compared to the change in the stockpile. Evidently, the stockpile policy seems to be destabilizing. However, except for wider fluctuations in the second year, which may have been caused by some speculative activities invoked in the rest of the world because of the change in the stockpile the rise in the LME price is smaller than the rise in the U.S. price, which indicates some degree of failure in achieving the objective of protecting the domestic industry, as well.

In the second case, 4(b), when the U.S. government is assumed not to intervene in the zinc market, one does not observe instability in the market. Prices, after an initial fall due to an increased supply, rise monotonically for a few years because of lagged adjustments and false expectations of the producers about the cessation of stockpile disposal program in the very near future. However, once it is realised that the expectations were false, the system settles down at the lower prices by the end of the simulation period. Further, it may be noted that, after the initial fall in prices, free-market price is observed to be higher than the U.S. price. This may be used to indicate that the U.S. industry is protected in this case. The same phenomenon is also reflected in a decrease in imports into the United States.

Table 7–3
Multiplier Simulations: An Increase in the Price of Aluminum by the International Bauxite Association (Model 2)

Year	*UCN*	*JCN*	*KCN*	*GCN*	*FCN*	*R1CN*	*R2CN*	*USTK*	*RSTK*	*LMPZ*	*USPZ*	*UIMP*
1965	100.00	100.00	100.00	100.00	100.00	100.00	100.00	100.00	100.00	100.00	100.00	100.00
1966	100.27	100.00	100.38	100.35	100.00	100.34	101.01	100.00	100.00	100.13	100.05	99.99
1967	100.45	100.00	100.52	100.74	100.28	100.69	101.99	100.00	100.33	100.34	100.14	99.98
1968	100.43	100.00	100.45	100.71	100.65	101.06	103.01	100.00	100.70	100.67	100.29	99.96
1969	100.12	100.00	100.13	100.86	101.10	100.90	102.55	100.00	101.41	100.47	100.22	99.97
1970	99.64	100.00	100.00	100.84	101.64	100.37	101.04	100.00	100.83	100.35	100.14	99.98
1971	99.83	100.00	100.00	100.83	100.00	100.00	100.00	100.00	100.57	100.21	100.07	99.99
1972	98.80	100.00	100.00	100.34	100.00	100.00	100.00	100.00	100.31	100.07	100.02	100.00
1973	99.80	100.00	100.00	100.22	100.00	100.00	100.00	100.00	100.10	100.02	100.01	100.00
1974	99.85	100.00	100.00	100.00	100.00	100.00	100.00	100.00	100.03	100.00	100.00	100.00

Note: Data are (Multiplier Solution/Control Solution) × 100.0

Table 7–4
Multiplier Simulations: A Change in the U.S. Government Stockpile Policy (Model 1)

	Assumption 4a.					*Assumption 4b.*				
Year	*USTK*	*RSTK*	*USPZ*	*LMPZ*	*UIMP*	*USTK*	*RSTK*	*USPZ*	*LMPZ*	*UIMP*
1965	100.00	100.00	100.00	100.00	100.00	147.55	100.00	92.41	80.00	101.70
1966	95.01	100.00	97.69	94.14	100.35	82.13	30.52	90.41	76.88	101.55
1967	100.00	85.20	101.24	102.80	99.85	91.34	64.77	98.75	95.43	100.32
1968	100.00	105.77	99.91	99.77	100.01	120.06	90.60	113.78	133.57	98.39
1969	100.00	99.53	100.01	100.00	100.00	148.43	170.10	118.60	145.05	97.77
1970	100.00	100.00	100.01	100.00	100.00	151.86	179.34	126.94	172.68	96.64
1971	100.00	100.00	100.02	100.00	100.00	157.01	220.47	139.02	194.31	95.76
1972	100.00	100.00	100.01	100.00	100.00	542.45	242.42	152.25	231.71	95.05
1973	100.00	100.00	100.00	100.00	100.00	413.20	285.64	141.04	169.06	96.97
1974	100.00	100.00	100.00	100.00	100.00	170.92	182.30	110.59	119.21	98.64

Note: (Multiplier Solution/Control Solution) × 100.0.

Notes

1. For a detailed discussion of both the problems involved—an analytical solution in the complex dynamic nonlinear systems and the computer simulation methodology—see Naylor (1971).

2. As a matter of fact, the aims of the policy have never been made public in precise terms. However, one can easily get impressions about these objectives from the published literature on the zinc industry. For details, see chapter 2.

3. See the discussion in chapter 2, and appendix to chapter 2.

8 Summary and Conclusions

The major aim of this study was to build an econometric model of the world zinc industry based on an adequate knowledge of its structural, behavioral, and organizational characteristics. It is expected that an in-depth study of these features will contribute towards the understanding of the zinc industry for all those concerned with the industry generally, and for economists interested in industry analysis or commodity studies, in particular. The econometric model of the industry is intended to help policymakers in formulating and evaluating certain important policies or in forecasting the major market variables. Besides, the model can be used to study the transmission of external influences on a national economy where the industry does not constitute an insignificant part of the economy. In the following four sections,a summary of the major findings of the study and some suggestions for further research in this area are reported in the following sequence: (1) organizational structure of the world zinc market, (2) models and results of estimation, (3) test of performance and applications, and (4) concluding remarks on further work in this area.

Organizational Structure

Consumption of zinc, an intermediate input widely used in construction, automobiles, arms and ammunition, household appliances, and many other manufactured commodities, is concentrated in the industrially advanced countries. The United States alone consumes about one-third of the total zinc used in the FME (Free Market Economies) world. Other major consumers of zinc are Japan, the United Kingdom, France, and West Germany. These five countries consume about seventy percent of the total zinc available in the FME world. However, this degree of concentration in consumption does not exert any significant influence in terms of market power on the buyer's side. This is so because of a large number of small and uncoordinated decision-making units that use zinc in numerous forms in manufacturing a wide variety of commodities.

Production of zinc ore and the associated mineral resources, though spread throughout the world, are more centralized in Canada, the USSR, the United States, Australia, Mexico, Peru, and, to a smaller degree, in a few of the European countries. Canada, Australia, Mexico, and Peru together produced about 53 percent of the FME world zinc ore production in 1974.

However, in terms of the international market for zinc, these countries, in the same year, shared in more than 80 percent of the exports of zinc ore and about 56 percent of the exports of zinc metal in the FME world. Apparently, this implies a high degree of concentration and, therefore, the possible presence of monopolistic elements on the sellers' side of the market. However, further investigations into the organizational structure of the industry, both at present and in the past, do not support this view. The basic arguments in this regard, as developed in this study, are briefly summarized below:

1. There are many producers in each of the above-mentioned countries whose decisions are not coordinated within each country (except for the United States). This means that a small number of countries does not imply a similarly small number of decision-making units in the market. As a matter of fact, the major producers of zinc are corporate groups. Some of these corporate groups operate across national boundaries. Thus, a closer look at the corporate structure of the world zinc industry is necessary.

A detailed investigation of this aspect reveals that in the year 1974, there were twenty-four corporate groups that had controlling interests in about 65 percent of the FME world mine capacity. The shares of the eleven largest and the four largest corporate groups (including their multinational operations) in the same year were 55 percent and 32 percent, respectively. Further, two of the largest four firms were Canadian and the other two were American. None of these firms controlled more than 10 percent of the FME world mine capacity. This degree of concentration, coupled with some other evidence in the literature against the existence of interdependent market behavior among producers, does not seem to provide adequate justification to hypothesize noncompetitive behavior in the world zinc industry.

2. Under certain circumstances, vertical integration is considered an important parameter in market behavior. Although a strong move towards vertical integration in the zinc industry is expected in the future, many large mining companies at present either toll smelt or sell a substantial part of their ores to smelters that are controlled by other large corporations. The European, U.S., and Japanese companies control 70 percent of the smelting capacity in the FME world, and this may have weakened the monopolistic power, if any, possessed by the mining corporations.

3. The major producers' moves for cartelization in the zinc industry during the interwar period, when the industry was even more concentrated than at present, do not suggest any optimism in this area. As a matter of fact, none of the cartels formed during the interwar period survived longer than a year.

However, during the last twenty-five years, some national and international organizations, notably the U.S. government and the United Nations, have ventured to influence the world zinc market. Though the efforts of the United States have been limited to the provision of statistical information with regard to the major market variables, the U.S. government has been observed to directly intervene in the working of the market forces to protect the domestic industry. Tariffs, subsidies, quotas, and stockpiles of zinc have been used to achieve these objectives. The secondary, though an important, effect of this intervention was that it gave an opportunity to the major producers—the four largest producers in the United States who controlled about 85 percent of the local mine and smelter production—to act together to achieve their objective of price stabilization. This type of concerted effort on the part of producers, or the patronage of such national policies, have not been observed elsewhere in the world zinc market.

Models and Results of Estimation

Given the above organizational structure of the zinc industry, the market form of econometric model is considered a suitable framework for an analysis of its structure, behavior, and performance. These models can also easily be used for policy formulations and forecasts. One of the basic characteristics of these models is that they contain a set of relationships pertaining to the demand for a commodity, its supply, and in some cases, inventories. The price of the commodity influences and is influenced by these variables.

In view of this modeling technique and the organizational structure of the world zinc industry, the econometric model of the market for zinc is divided into two subsystems: (1) the free-market world outside the United States, where competitive market behavior is assumed, and (2) the market for zinc within the United States, where some elements of noncompetitive behavior, as discussed above, are incorporated. Both the subsystems are linked together through prices, interregional trade, and exchange rates. A brief description of the basic relationships in the model and the results of their estimation are given below.

Free Market World (Excluding the United States)

Demand. Total consumer demand for zinc is divided into six regions: Japan, the United Kingdom, West Germany, France, the rest of the developed world, and the rest of the world. This subdivision aims to provide reasonable scope for incorporating structural differences in demand patterns in different regions. There are also some structural differences in the demand patterns according to the sectors of demand within each region. These are included in the second version of the model through a subdivision of the total consumer demand for zinc in each major consumer country into six categories (zinc used for galvanizing, die casts, brass, rolled zinc, zinc oxides, and a miscellaneous category). Demand for zinc in each category of consumption in each major country is hypothesized to be influenced by the price of zinc, prices of substitutes and complements, and the relevant activity variables. The specification is consistent with the hypothesis of cost minimization behavior of the consumers. Prices of zinc and of the substitutes are assumed to follow and inverted V-shape polynomial lag structure (the most successful lag structure found in the estimation of these demand relationships).

As many of the uses of zinc are specific to its technical properties, the responses of consumption to prices of zinc and its substitutes are generally poor. In no case were the coefficients of current price variables meaningful or statistically significant. In general, the response of consumption to prices starts after the lapse of a year or two. In the aggregative version of the model, the long-run price elasticities of demand vary from −0.04 for Japan to −0.78 for West Germany.

The elasticity estimates are considerably improved in the second version of the model, where the demand equations for each of the major consumer countries are disaggregated according to the six sectors of demand for zinc. The average weighted demand elasticity estimates (consumption shares of the sectors in each country are taken as weights) for Japan, West Germany, the U.K., and France are −0.23, −1.22, −0.29, and −0.69, respectively. In general, the uses of zinc for die casts, brass, and rolled zinc are more price-elastic than those for galvanizing and oxides. The elasticity estimates for the rest of the world are close to zero.

Supply. The total supply of zinc is divided into primary supply and secondary supply (zinc recovered from scrap). Primary supply, in turn, is subdivided according to six major producer areas (Australia, Canada, Mexico, Peru, Europe, and the rest of the world) to account for the structural and institutional differences in the regions.

The major variables explaining the supply of primary zinc are: (1) the price of zinc, (2) prices of coproducts, (3) the level of mine capacity in the area concerned, (4) wages in mining, and (5) a time trend to capture the long-run influences of technological changes. Average variable costs as reflected by wages are not found statistically significant except for Canada and Peru. The prices of coproducts were significant only for Europe.

The producers' response to the price of zinc is based on the partial-adjustment hypothesis. In most cases, the current-year price elasticity is close to zero or is wrongly signed, which is expected in view of the technological lags involved. The long-run elasticities of supply are also fairly low, ranging from 0.24 for Australia to 0.62 for Peru. The estimated value of these elasticities for Canada, Mexico, Europe, and "Rest of the World" (excluding the United States) are 0.45, 0.28, 0.32, and 0.57, respectively. Higher elasticity values for Peru, Canada, and "Rest-of-the-World" indicate newer mine deposits in those countries. The same is reflected in a response to the capacity variable in these countries as compared with Australia, Mexico, and Europe where the mine deposits are relatively old.

The supply of zinc from secondary sources is assumed to compete with the available primary resources. The ratio of consumption to primary resources and a time-trend variable to reflect accumulation of old scrap over time and the changes in technology of recovery, are found very significant in explaining supply from secondary sources.

Price and Inventory. The price of zinc in the free-market world (excluding the United States) is assumed to be sensitive to the demand and supply forces of the market. The price behavior in this market, therefore, is hypothesized to depend on the ratio of stocks to the level of demand. The other variables included in the equation are the price of zinc lagged by one year (to capture the lagged effects) and some exogenous policy variables influencing this market. In particular, the U.S. government stockpiles and a dummy variable for the quota period in the United States are also included. The estimated equation indicates a 50 percent rise in the price for a 1 percent rise in the stock demand ratio. The price is also very sensitive to the fluctuations in the U.S. government's stockpile policy.

The U.S. Subsystem

The demand and supply forces in this subsystem are assumed to be determined in the same way as for the other subsystem, the only difference being the determination of secondary supply. The secondary supply in the United States is subdivided according to two sources—old scrap and new scrap. The lack of adequate data did not permit this division in the other subsystem.

Demand. Demand for zinc in the United States is found to be more sensitive to price (elasticity = −0.98) as compared to that in any other country in the non-U.S. subsystem. This long-run elasticity value increases to −1.17 when the consumption was disaggregated according to the six major categories of demand. Within the six categories of demand, the use of zinc for die casts recorded the highest sensitivity to price (elasticity = −2.06). Variations in the price of substitute materials do not appear to influence the major categories of zinc consumption substantially.

Supply. The supply of primary zinc in the United States is more sensitive to the price of zinc as compared to the producers' response to price in other areas. The estimated value of price elasticity of supply for the United States is 0.85. The elasticity of supply with respect to the capacity variable is 0.76.

Prices of coproducts and wages, though important in explaining variations in supply, are statistically not significant.

Secondary supply is modeled according to whether it is recovered from old scrap or new scrap. Supply of zinc from new scrap depends on the source of the new scrap; that is, on the level of metal fabricated for consumption, and the price of zinc. However, the supply of zinc from old scrap depends mainly on the availability of primary resources in relation to the level of consumption, the accumulation of old scrap over time, and the technology of recovery. In the cases of both new and old scrap, resource variables are found more powerful as compared to the price variable in explaining the supply from these sources.

Price and Inventory. The U.S.-produced price of zinc is viewed, at least in the short run, as the price administered by the U.S. producers through variations in their stock holdings (in relation to the level of consumption) and the capacity utilization ratio. However, in the long run, the U.S. price, in general, is assumed to respond to the forces of the world demand and supply as indicated by the free-market price.

The estimated U.S. price equation indicates a predominance of the capacity utilization ratio over the variations in the stock consumption variable in explaining the U.S. price movements. The long-run tendency of the U.S. price, however, is well captured by the free-market price itself. The elasticity estimates of the U.S. price with respect to these three variables are −0.85, −0.03, and 0.51, respectively.

Trade. Trade between the United States and the non-U.S. free-market world is hypothesized as a behavioral equation in terms of the U.S. import demand function. The U.S. demand for imports is assumed to be influenced by the differential between the free-market and the U.S. price, the activity level in the United States, and the pound-sterling/U.S.-dollar exchange rate. The estimated value of elasticity of import demand with respect to income, price variables, and exchange rate are 0.48, −0.12, and 0.56, respectively.

Besides interregional trade, the two stock identities, one each for the U.S. and the non-U.S. world, serve to close the system.

Test of Performance and Applications

Test of Performance

The estimated structure of the model for both subsystems together is tested for its predictive ability based on the method of sample-period dynamic simulations. Results of these dynamic simulations are assessed in terms of the average percentage absolute deviations (APAD) defined as the average of the absolute deviations of the solution values from the actual values which, in turn, are expressed as a percentage of the actual values. The summary measure is intended to measure the failure of the model to reproduce the historical data. A second summary measure (APD) in terms of the algebraic rather than absolute deviations, defined similarly to the above measure, is also used to reveal any systematic tendency of the model to underpredict or overpredict all the sample values.

Both structures of the model perform in a reasonably satisfactory manner. Equations relating to stocks and prices produce larger prediction errors as compared to the equations for demand, supply, and interregional trade. Average percentage absolute errors in the case of demand equations are generally between 2.0 and 4.0, with an exception of 4.5 for West Germany. In the case of primary supply equations, the errors are found to be between 2 and 6 percent except for Canada where they are on the order of 10 percent. The algebraic percentage error for Canadian supply, which is also −10, reveals that the model has systematically underestimated the historical data. Besides many new mine discoveries in Canada during the simulation period, this large error may be the result of some specification errors in the Canadian supply equation. The larger errors in predicting stocks (17 for the United States and 39 for the rest of the world) and prices (5 for the United States and 12 for the rest of the world) may be attributed to (1) inaccuracy of data for stocks (U.S. stocks and prices have much smaller errors where more accurate data for stocks are available), and (2) frequent short-term fluctuations in stocks and prices as revealed by a comparatively lower figure in terms of the average percentage algebraic errors (−2.7 for U.S. stocks and 3.0 for rest-of-the-world stocks, −0.95 for U.S. price and −0.24 for rest-of-the-world price). The average percentage absolute errors and average percentage algebric errors for trade are 5.0 and −0.6, respectively. Time charts of the actual and simulated values of all the major endogenous variables of the model are given in the appendix.

Applications

Given a reasonably satisfactory predictive ability of the model, the model is used to explore performance properties of the world zinc market for some exogenously given short-run and long-run disturbances. Dynamic multiplier simulations caused by the disturbances are compared with the sample period dynamic simulations discussed above. Results of four such experiments are summarized below.

1. The first experiment focuses on the stability properties of the market for (a) a temporary shift in demand due to a 1-percent rise in the activity level

in the beginning year (1965) of the simulation period, and (b) a long-term shift in demand due to a continued increase in the activity level by 1 percent. In case (a) the system returns to the base solution by the fourth year. The number of years taken to stabilize the price indicates delayed responses in the system. In case (b) the price of zinc in both the United States and the rest of the world rise by an average of one-third of 1 percent above the control solution; and this increase continues until the end of the simulation period.

2. In the second experiment an attempt is made to look into the implications of a technological change in the automobile sector, one of the largest consumers of zinc. The disturbance is introduced through a reduction of the estimated coefficient of the die casting sector by 33 percent. As a result, in most countries the fall in consumption is recorded as between 10 and 30 percent as compared to the control solutions. Fall in demand in relation to supply resulted in a fall in both the free-market and the U.S. prices by an average of 12 and 25 percent, respectively. This suggests that the world zinc industry, in the future, should be paying a great deal of attention in its research activities to finding new avenues for the use of zinc to insulate it from such possible changes.

3. In the third experiment, the price of aluminum, the major substitute for zinc, is increased by 10 percent in the beginning year of the simulation period. In general, the results are to increase the consumption of zinc and to raise the price of zinc. However, the consequent increases in the price of zinc are very small (less than 1 percent). This reflects a very weak substitutability in the system.

4. The fourth experiment concerns an evaluation of the stockpile policy of the U.S. government. The major aims of the stockpile policy throughout the period have been protection of the domestic industry and removal of short-run fluctuations in the price of zinc. In part (a) of this experiment, an attempt is made to test the validity of these objectives through a temporary increase (1 percent) in the stockpile policy. In part (b), the likely nature of the market performance is investigated in a hypothetical situation where the U.S. government does not intervene in the market and, rather, is prepared to release all stockpiles evenly over fourteen years beginning in 1963 (by 100,000 tons per year).

In (a), the results of the stockpile policy do not reveal the fulfillment of the objectives. Rather, the price of zinc fluctuates and often the U.S. price was higher than the world market price. In case (b), the results indicate less instability in the prices and also a larger increase in the free-market price as compared to the U.S. price. This implies that the stockpile policy of the U.S. government was either not properly geared to the objective, or was an unnecessary intervention in the world zinc market.

Concluding Remarks

The present study is the first attempt to investigate systematically all the institutional and market forces underlying the world zinc industry. These investigations provide a framework within which a detailed market form of econometric model is developed and estimated. A reasonably satisfactory performance of the model in a dynamic context demonstrates that it is a useful tool for policy formulations and forecasts. However, the study can be claimed only to provide groundwork that can be used to extend the analysis in many directions. In particular, for a more successful policy evaluation, one may, in the future, extend the analysis in the following ways:

1. by incorporating the technological information on mining and smelting;
2. by linking production, capacity, and resources; and
3. by including more accurate inventory behvior and the associated expectations mechanisms.

A lack of adequate information on the major economic variables relating to the technologies of mining and smelting has precluded a fuller treatment of the supply side of the world zinc market. In the future, more detailed information on the relevant variables of production technology can be expected to provide larger scope for an econometric or even a linear-programming analysis of the supply side. A more detailed analysis of policies, more particularly in terms of direction of flows of zinc ore and metal, might easily be carried out if the supply side, based on the programming approach, were integrated to the demand side as dealt with in this study.

The available data on resources containing zinc are both inadequate and inaccurate for a meaningful long-term analysis of the market. In general, the data on resources are a result of some educated guesswork at each point in time. This type of data, even if it may be taken as the most accurate available, reflects only stocks or inventories of resources at each point in time. What is needed is the data on resources which would properly incorporate the long-run developments in the resources through explorations and mine developments. Given this type of data on resources, and some more data on capacity variables, one could extend the analysis for more successful policy formulations and forecasts. The problems are similar with respect to the data on inventory holdings of the dealers and producers, which has precluded modeling of a more accurate expectations mechanism.

Thus, there is a good deal of scope for further research work in this area, depending on the availability of more adequate and accurate information. However, in the meantime, it is expected that this study will contribute to the present understanding of the structure, behavior, and performance of the world zinc industry, will provide some help to the policy makers and planners concerned with this industry, and will stimulate more interest in the research work in this area.

Appendix: Plots of Dynamic Simulations (Model 2)

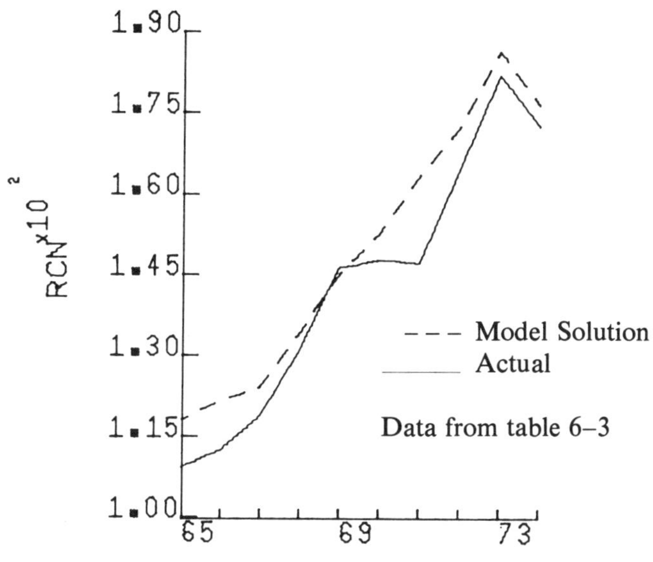

Figure A–1. FME World (except U.S.) Zinc Consumption versus Time

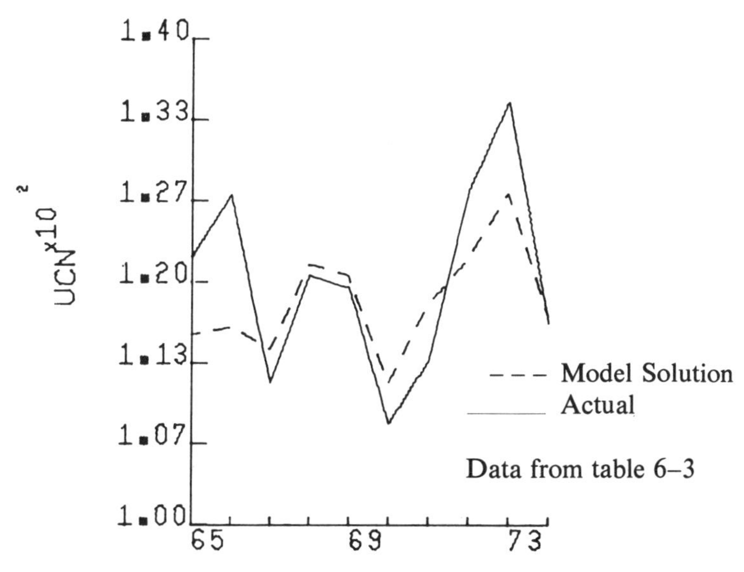

Figure A–2. U.S. Zinc Consumption versus Time

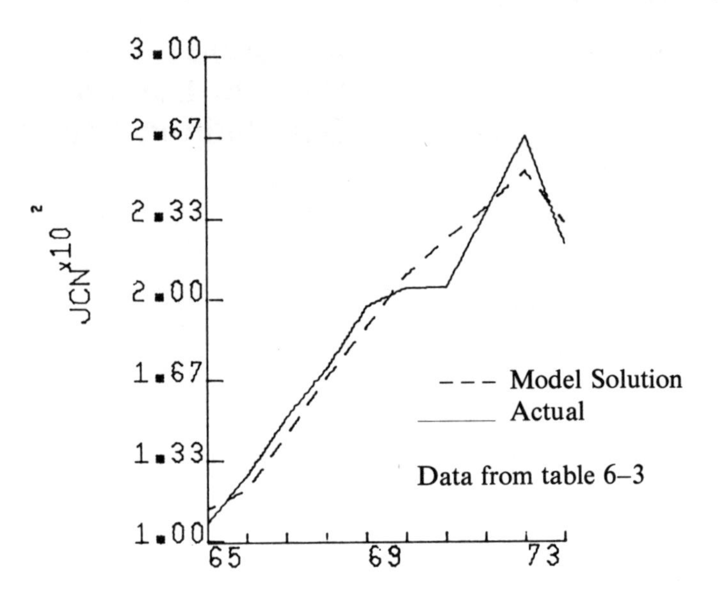

Figure A–3. Japan Zinc Consumption versus Time

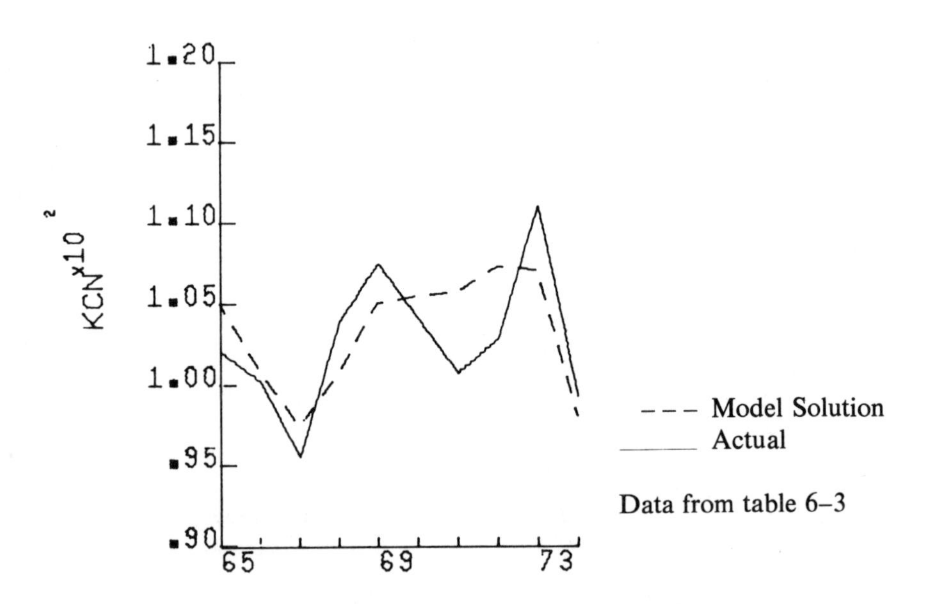

Figure A–4. U.K. Zinc Consumption versus Time

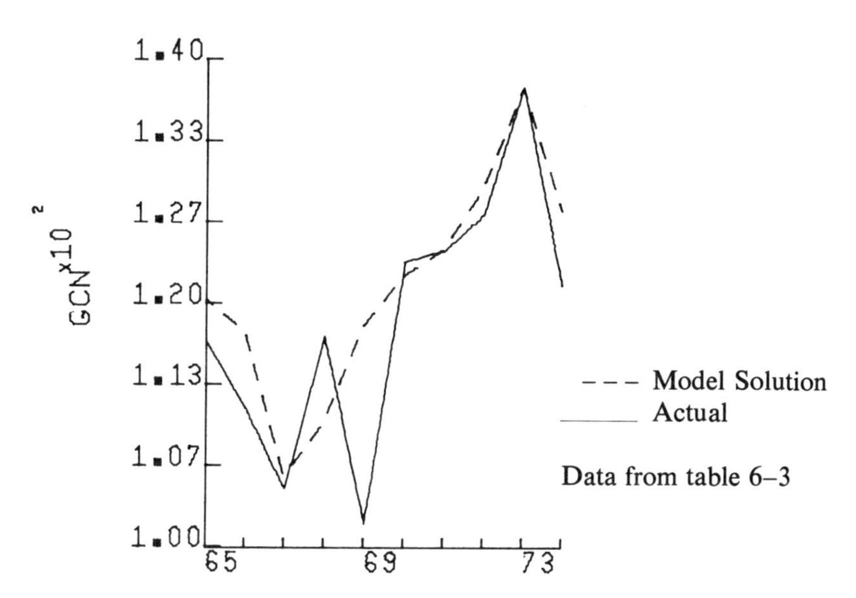

Figure A–5. West Germany Zinc Consumption versus Time

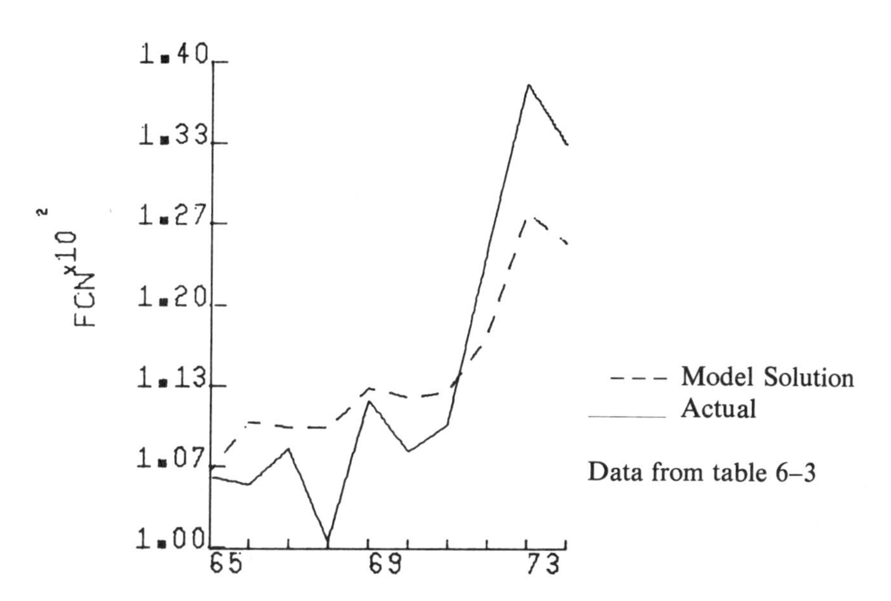

Figure A–6. France Zinc Consumption versus Time

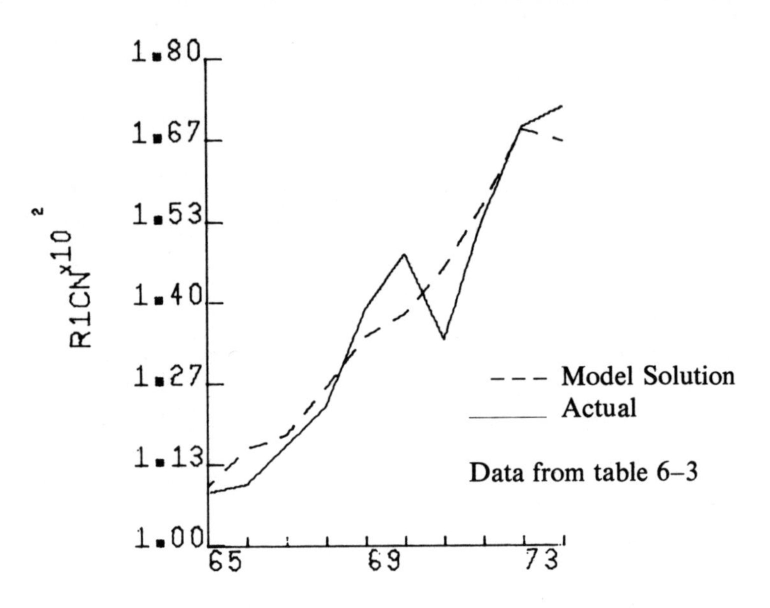

Figure A–7. Rest of Developed World Zinc Consumption versus Time

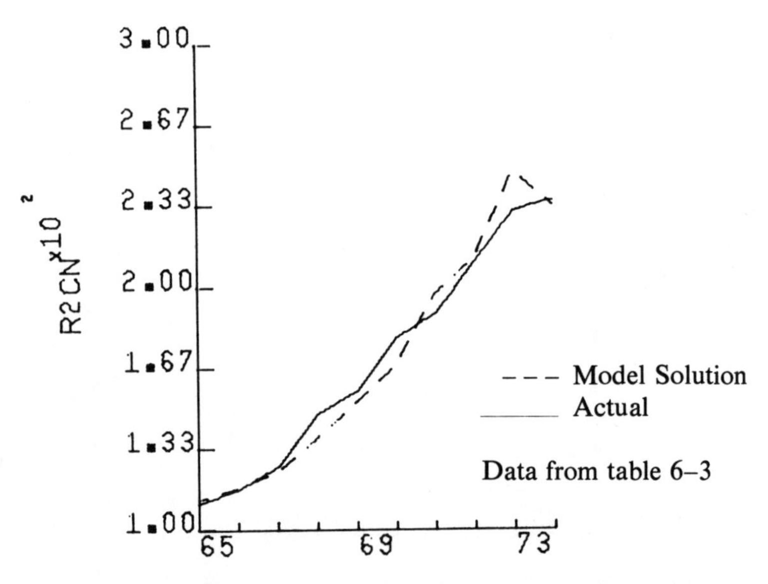

Figure A–8. Rest of World Zinc Consumption versus Time

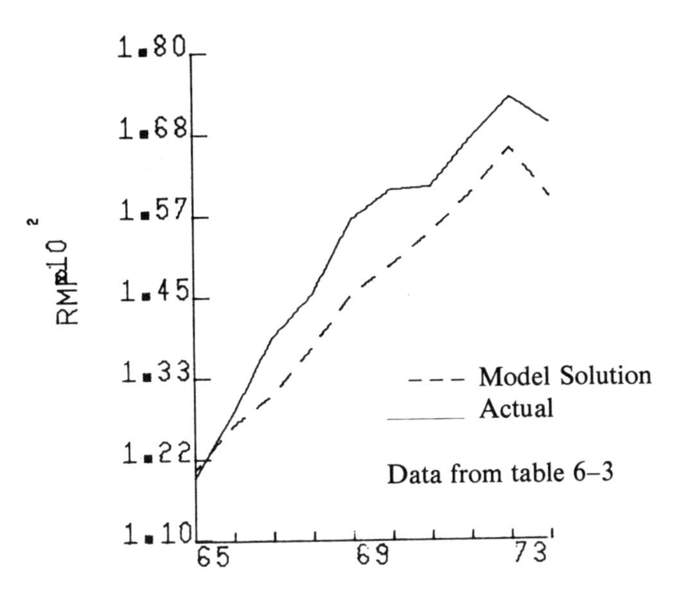

Figure A–9. FME World (except U.S.) Zinc Mine Production versus Time

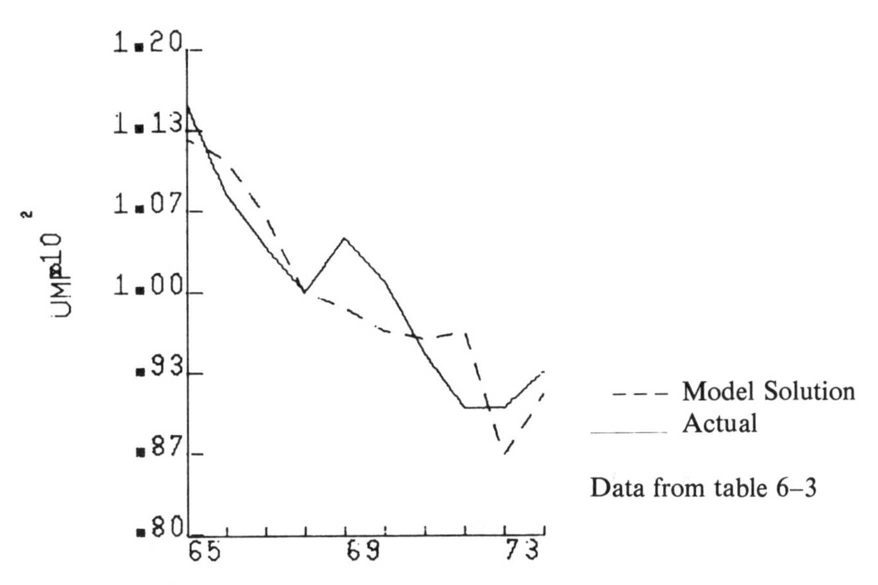

Figure A–10. U.S. Zinc Mine Production versus Time

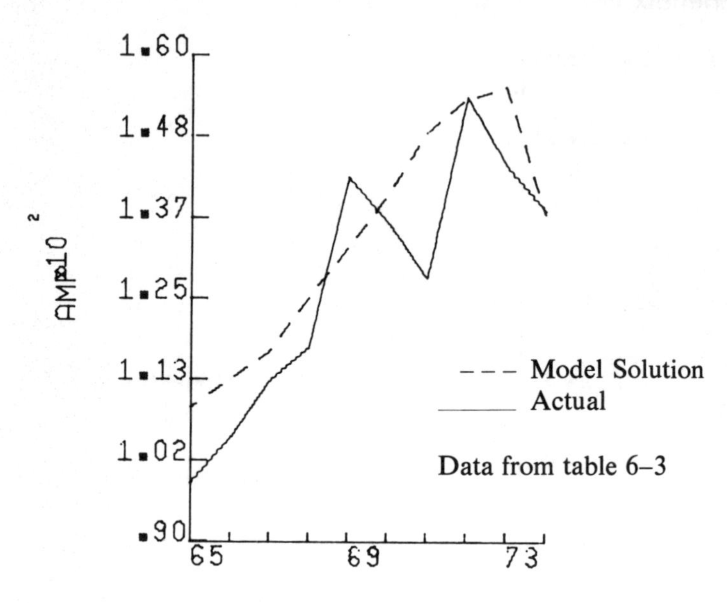

Figure A–11. Australia Zinc Mine Production versus Time

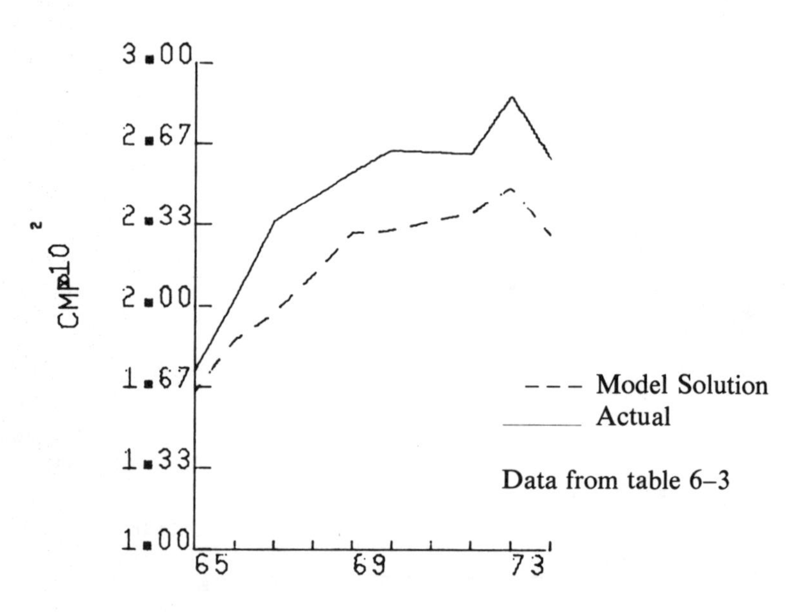

Figure A–12. Canada Zinc Mine Production versus Time

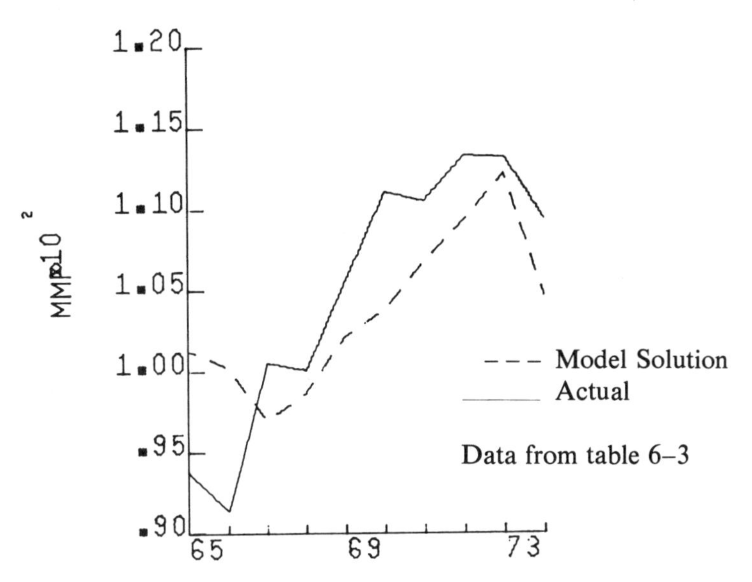

Figure A–13. Mexico Zinc Mine Production versus Time

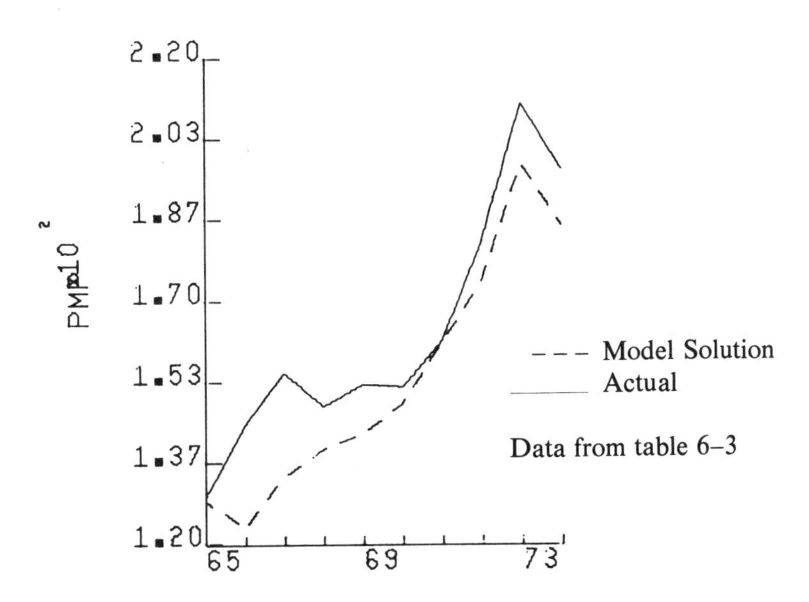

Figure A–14. Peru Zinc Mine Production versus Time

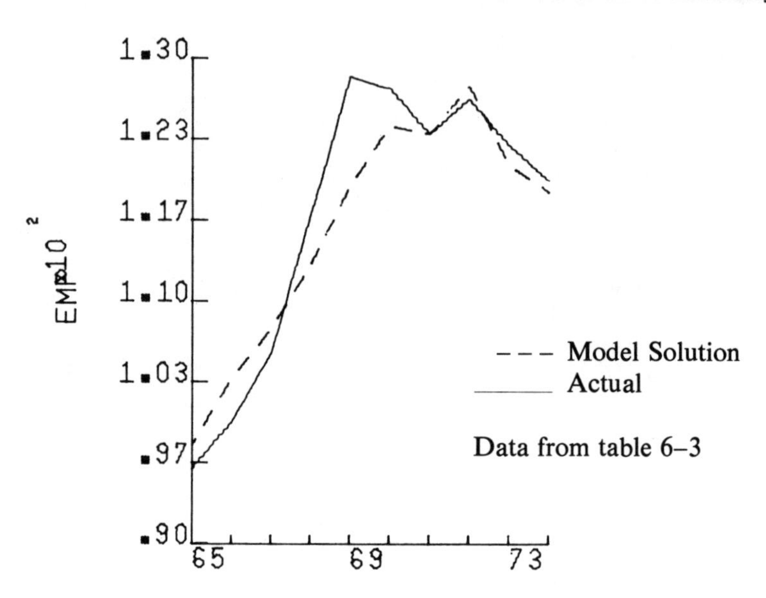

Figure A–15. Europe Zinc Mine Production versus Time

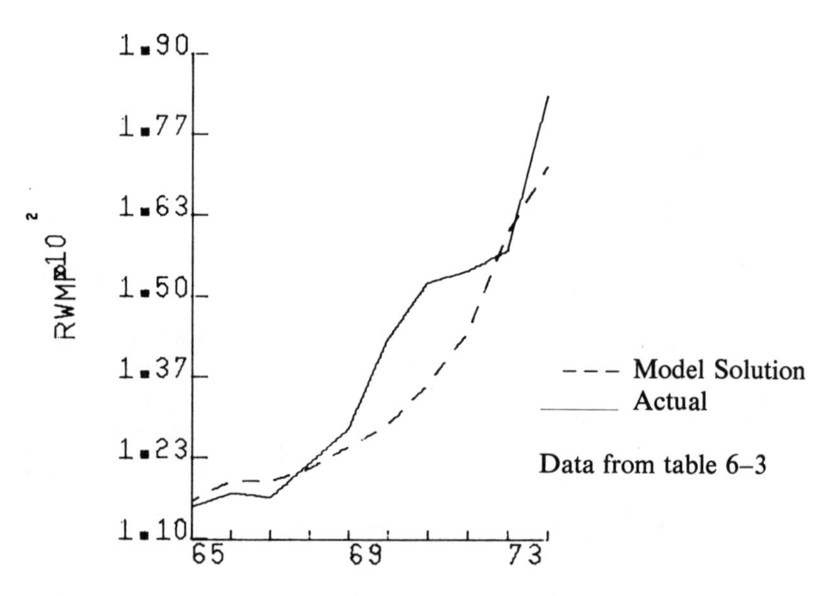

Figure A–16. Rest of World Zinc Mine Production versus Time

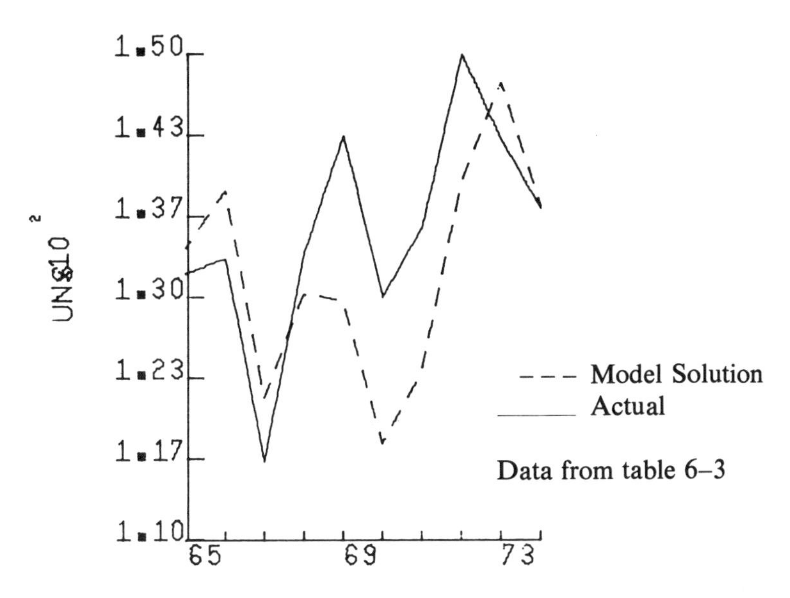

Figure A–17. U.S. Production of Zinc from New Scrap versus Time

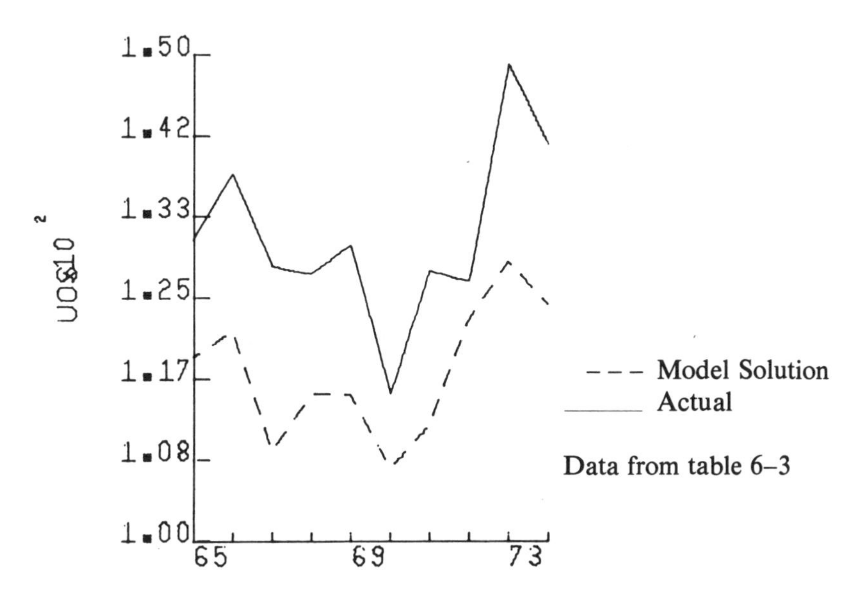

Figure A–18. U.S. Production of Zinc from Old Scrap versus Time

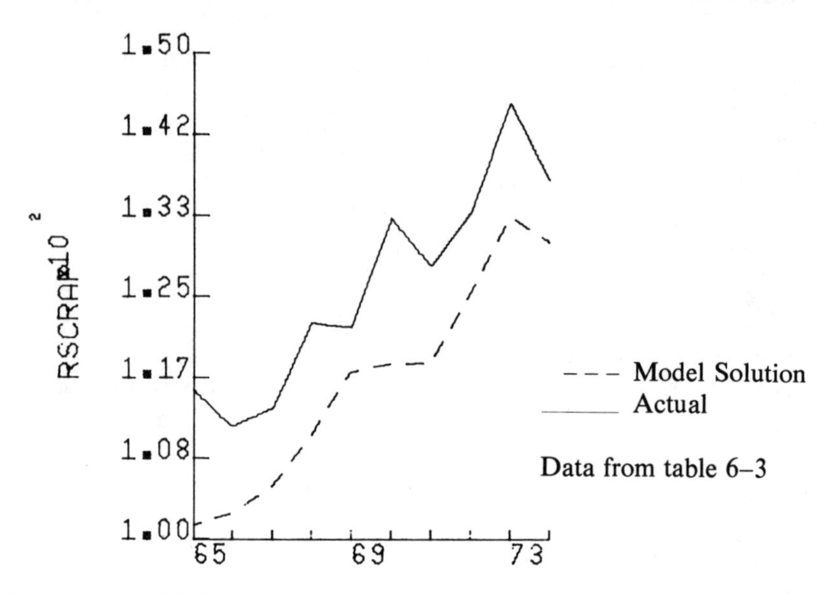

Figure A–19. FME World (except U.S.) Production of Zinc from Scrap versus Time

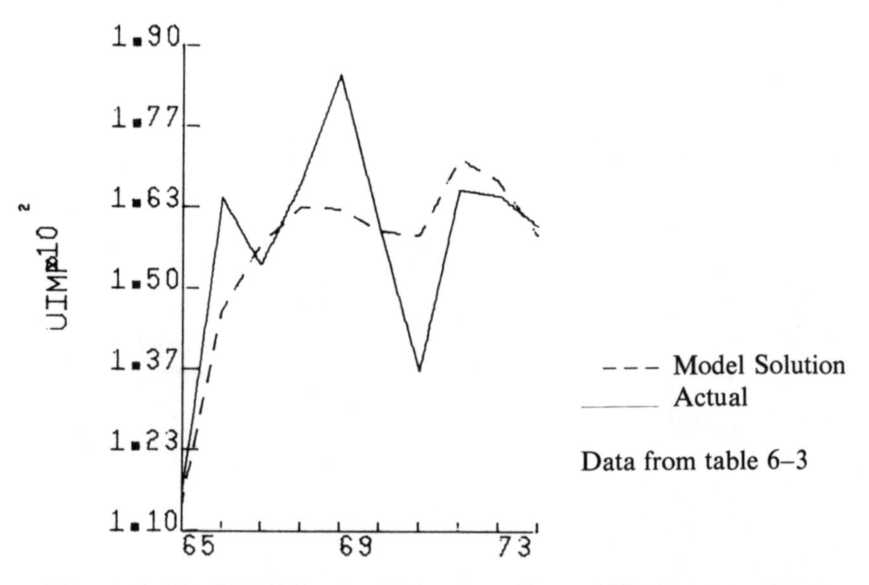

Figure A–20. U.S. Imports of Zinc from Rest of World versus Time

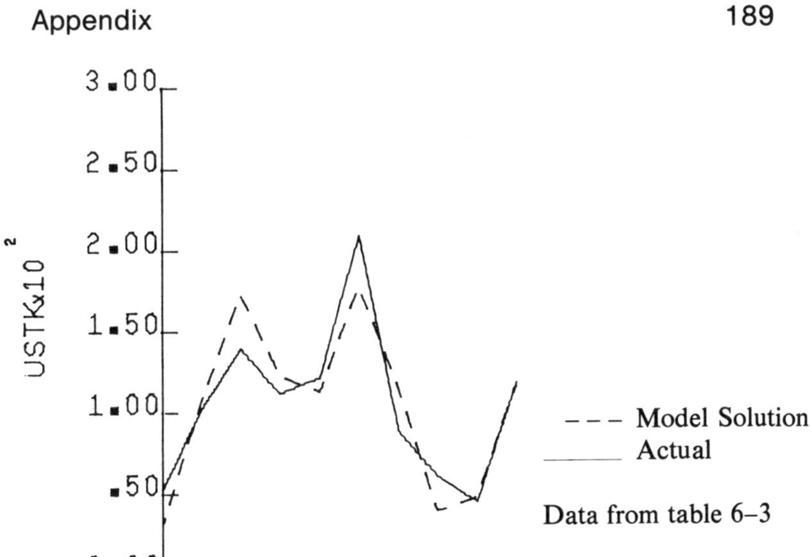

Figure A–21. U.S. Stocks of Zinc versus Time

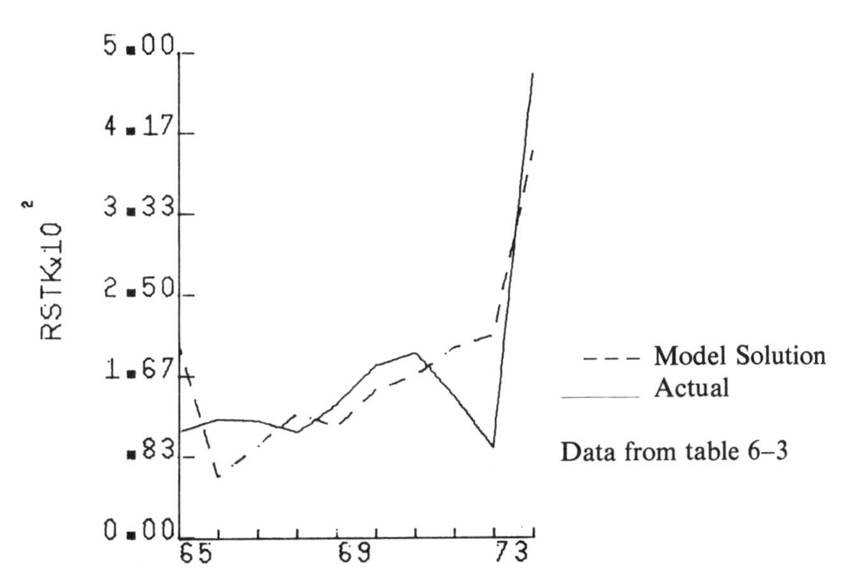

Figure A–22. Rest of World Stocks of Zinc versus Time

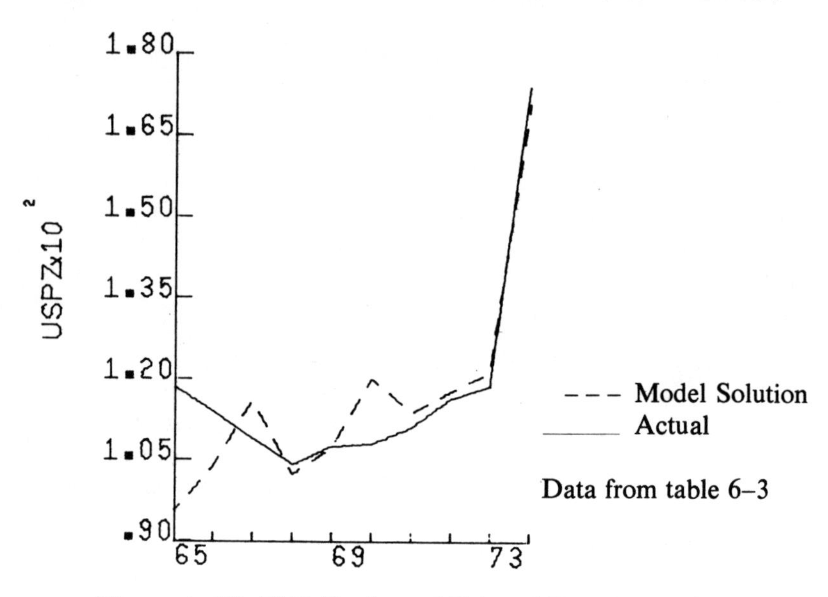

Figure A–23. U.S. Producers' Price of Zinc versus Time

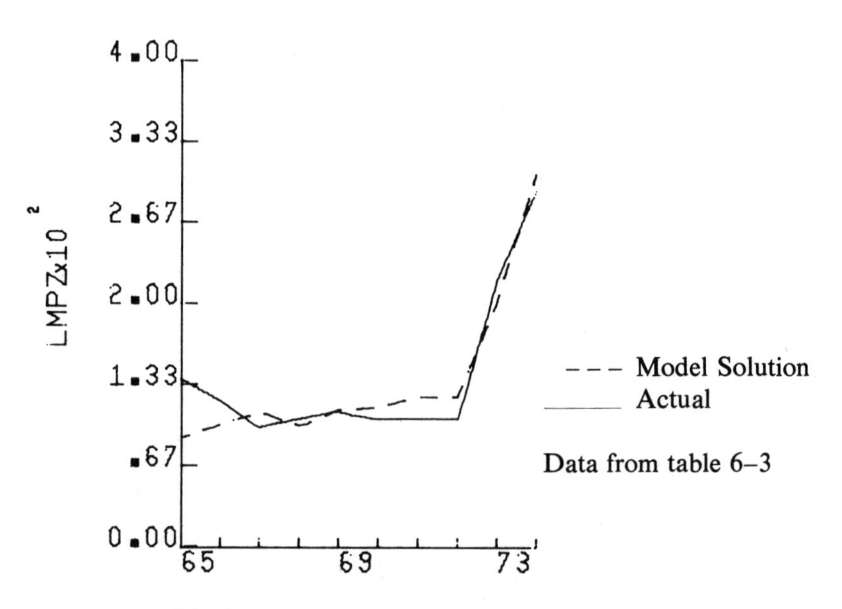

Figure A–24. London Metal Exchange Price of Zinc versus Time

Bibliography

Adams, F.G. 1972. "The Impact of Cobalt Production from the Ocean Floor: A Review of Present Empirical Knowledge and Preliminary Appraisal." Paper prepared for UNCTAD, 15 March, *Philadelphia: Economic Research Unit*, University of Pennsylvania.

Adams, F.G. 1973. "The Impact of Copper Production from the Ocean Floor: Application of an Econometric Model." Paper prepared for UNCTAD, December, *Philadelphia: Economic Research Unit*, University of Pennsylvania.

Adams, F.G. 1973a. "The Integration of the World Primary Commodity Markets into Project Link: The Example of Copper." Paper prepared for the annual meeting of Project LINK, Stockholm, September, *Philadelphia: Economics Research Unit*, University of Pennsylvania.

Adams, F.G., and Behrman, J.R. 1977. *Econometric Models of World Agricultural Commodity Markets: Cocoa, Coffee, Tea, Wool, Cotton, Sugar, Wheat, Rice*. Cambridge, Mass.: Ballinger.

Adelman, I., and Adelman, F. 1959. "The Dynamic Properties of the Klein-Goldberger Model." *Econometrica* (October): 596–625.

Almon, S. 1965. "The Distributed Lag Between Capital Appropriations and Net Expenditures." *Econometrica* (January): 178–196.

American Bureau of Metal Statistics. *Yearbooks*. New York.

American Metal Market. *Metal Statistics* (annual). New York: Fairchild Publications.

Andrews, P.W. 1970. *The U.S. Zinc Industry under the Import Quota Program*. Ottawa: Department of Energy, Mines and Resources.

Avramides, A., and Cross, J.S. 1973. "NPC Analysis of Oil and Gas Supply." In *Working Paper EN-1*, edited by M.F. Searl. Washington: Resources for the Future, Inc.

Ballmer, R.W. 1960. "Copper Market Fluctuations: An Industrial Dynamics Study." M.S. thesis, Massachusetts Institute of Technology.

Banks, F.E. 1971. "An Econometric Note on the Demand for Refined Zinc." *Zeitschrift für Nationalökonomi* 31: 443–452.

Banks, F.E. 1974. "The World Copper Market: An Economic Analysis." Cambridge, Mass.: Ballinger.

Behrman, J.R. 1972. "Forecasting Properties and Prototype Simulation of a Model of the Copper Market." Special Report. Philadelphia: Wharton Economic Forecasting Associates.

Behrman, J.R. 1976. "International Commodity Agreements." Mimeographed. Prepared as part of the Overseas Development Council, New International Economic Order Research Project, October, Philadelphia: Economic Research Unit, University of Pennsylvania.

Burrows, J.C. 1971. *Cobalt: An Industry Analysis.* Lexington, Mass.: Lexington Books, D.C. Heath and Company.

Cairns, J.H., and Gilbert, P.T. 1967. *The Technology of Heavy Non-Ferrous Metals and Alloys: Copper, Nickel, Zinc, Tin, Lead.* Cleveland: C.R.C. Press.

Cammarota, V.A.; Babitzke, H.R.; and Hague, J.H. 1975. "Zinc," *Mineral Facts and Problems*. Washington: U.S. Bureau of Mines.

Cochrane, D., and Orcutt, G.H. 1949. "An Application of Least Squares Regression to Relationships Containing Autocorrelated Error Terms." *Journal of the American Statistical Association* 44: 32–61.

Dayananda, M.D. 1977. "Potential Cartelisation in a Monopsonistic Market Structure: A Model of the World Tea Market with Special Reference to Sri Lanka." Ph.D. dissertation, McMaster University.

Department of Energy, Mines and Resources. 1976. *Zinc, Mineral Bulletin*, MR 159, Ottawa.

Desai, M. 1966. "An Econometric Model of the World Tin Economy, 1948–61." *Econometrica* 34: 105–134.

Eckbo, P.L. 1975. "OPEC and the Experience of Previous International Commodity Cartels." *Energy Laboratory Working Paper, Number 75–008 WP*. Cambridge, Mass.: Massachusetts Institute of Technology.

Edwards, J.B., and Orcutt, G.H. 1969. "Should Aggregation Prior to Estimation be the Rule?" *Review of Economics and Statistics* (November) 409–420.

Elliot, W.Y., et al. 1937. *International Control of the Non-Ferrous Metals.* New York: Macmillan.

Engineering and Mining Journal. Various issues. New York: McGraw Hill Inc.

Epps, Mary L.S. 1970. "A Computer Simulation Model of the World Coffee Economy." Ph.D. dissertation, Duke University.

Ertek, T. 1967. "World Demand for Copper, 1948–63: An Econometric Study." Ph.D. dissertation, University of Wisconsin.

Etherington D.M. 1972. "An International Tea Trade Policy for East Africa: An Exercise in Oligopolistic Reasoning." *Stanford Univeristy Food Research Institute Studies* 11:89–108.

Fisher, F.M.; Cootner, P.H.; and Baily, M. 1972. "An Econometric Model of the World Copper Industry." *Bell Journal of Economics and Management Science* no. 3: 568–609.

Friedman, M. 1952. "Comment" in *A Survey of Contemporary Economics.* Edited by B.F. Haley. Homewood, Ill: Richard D. Irwin Inc.

Friedman, M. 1953. *Essays in Positive Economics.* Chicago: University of Chicago Press.

Gordon, R.L. 1967. "A Reinterpretation of the Pure Theory of Exhaustion." *Journal of Political Economy* 75: 274–86.

Gupta, K.L. 1969. *Aggregation in Economics, A Theoretical and Empirical Study*. Rotterdam University Press.

Heindl, R.A. 1970. "Zinc," *Mineral Facts and Problems*. Washington: U.S. Bureau of Mines.

Herfindahl, O.C. 1955. "Some Fundamentals of Mineral Economics." *Land Economics* 31: 131–138.

Hotelling, H. 1931. "The Economics of Exhaustible Resources." *Journal of Political Economy* 39: 137–175.

International Lead and Zinc Study Group. 1966. *Lead and Zinc: Factors Affecting Consumption*. New York: United Nations.

International Lead and Zinc Study Group. *Lead and Zinc Statistics*. New York: United Nations.

Johnston, J. 1972. *Econometric Methods*. 2nd ed. New York: McGraw Hill.

Khanna, I. 1972. "Forecasting the Price of Copper." *Business Economist* no. 4. Herts: Society of Business Economics.

Koopmans, T.C. 1957. *Three Essays on the State of Economic Science*. New York: McGraw Hill.

Krenin, M.E., and Finger, J.M. 1976. "A New International Economic Order: A Critical Survey of its Issues." *Journal of World Trade Law* (September/October).

Labys, W.C. 1973. *Dynamic Commodity Models: Specification, Estimation and Simulation*. Lexington, Mass: Lexington Books, D.C. Heath and Company.

Labys, W.C., ed. 1975. *Quantitative Models of Commodity Markets*. Cambridge, Mass: Ballinger.

Levhari, D. and Liviation, N. 1977. "Notes on Hotelling's Economics of Exhaustible Resources." *Canadian Journal of Economics* (May) 177–192.

Mahalingsivam, R. 1969. "Market for Canadian Refined Copper: An Econometric Study." Ph.D. dissertation, University of Toronto.

Mathewson, C.H., ed. 1969. *Zinc: The Science and Technology of the Metal, Its Alloys and Compounds*. New York: Reinhold.

McMahon, A.D.; Cotterill, D.H.; Dunham, J.T.; and Rice, W.L. 1974. *The U.S. Zinc Industry: A Historical Perspective* (IC 8629). Washington: U.S. Bureau of Mines.

Metal Bulletin. Various issues. Survey: Metal Bulletin Ltd.

Metal Statictics. Frankfurt: Metallgesellschaft Aktiengesellschaft.

Metals Week. New York: McGraw Hill Inc.

Moody's Industrial Manual. 1976. New York: Moody's Investors Service Inc.

Mujeri, M.K. 1978. "The World Market for Jute: An Econometric Analysis." Ph.D. dissertation, McMaster University.

Naylor, T.H. 1971. *Computer Simulation Experiments with Models of Economic Systems*. New York: John Wiley & Sons, Inc.

Organization for Economic Cooperation and Development. *Main Economic Indicators*. (1955–1975). Paris.

Orcutt, G.H.; Watts, H.W.; and Edwards, J.B. 1968. "Data Aggregation and Information Loss." *American Economic Review* 773–787.

Peterson, F.M., and Fisher, A.C. 1977. "The Exploitation of Extractive Resources, A Survey." *The Economic Journal* (December) 681–721.

Pindyck, R.S. 1978. "Gains to Producers from the Cartelisation of Exhaustible Resources." *Review of Economics and Statistics* (May), 238–251.

Rachmalla, K.S., and Bell, D.H. 1976. *Towards a Zinc Policy for the Province of Ontario*. Mineral Policy Background Paper No. 3. Toronto: Ministry of Natural Resources.

Roskill Reports on Metals and Minerals. 1974. *The Economics of Zinc*. London: Roskill Information Services.

Scherer, F.M. 1970. *Industrial Market Structure and Economic Performance*. Chicago: Rand McNally & Co.

Solow, R.M. 1974. "The Economics of Resources or the Resources of Economics." Proceedings, *American Economic Review* 45: 1–14.

Stiglitz, J.E. 1976. "Monopoly and the Rate of Extraction of Exhaustible Resource." *American Economic Review*. September, 651–655.

Theil, H. 1954. *Linear Aggregation of Economic Relations*. Amsterdam: North Holland.

United Nations. *Monthly Bulletin of Statistics*. Various issues. New York.

United Nations. *International Labour Statistics*. 1955–1975 Geneva: International Labour Office.

United Nations. *Yearbook of Industrial Statistics*. 1955–1975.

United Nations. *International Financial Statistics*. Washington: International Monetary Fund.

UNCTAD 1974. "An Integrated Program for Commodities." (TD/b/c.1/166) Geneva.

U.S. Bureau of Mines. *Minerals Yearbook*. 1937–1975.

Wharton Econometric Forecasting Associates Inc. 1973. *Forecasts and Analysis of Zinc Market*. Study prepared for Property Management and Disposal Service, General Services Administration, Washington, D.C. (Jointly with Charles River Associates, Inc.)

World Mines Register. 1975. New York.

Index

Index

About the Author

Satyadev Gupta is currently a lecturer in economics at the University of the West Indies. He received his formal training at the Delhi School of Economics, India, and McMaster University, Canada. He has taught at the University of Delhi and was a visiting research scholar at the University of Pennsylvania. His research and publications include many articles in the fields of industrial economics, applied econometrics, and public economics.